Methods and Protocols in Food Science

Series Editor
Anderson S. Sant'Ana
University of Campinas
Campinas, Brazil

For further volumes:
http://www.springer.com/series/16556

Methods and Protocols in Food Science series is devoted to the publication of research protocols and methodologies in all fields of food science.

Volumes and chapters will be organized by field and presented in such way that the readers will be able to reproduce the experiments in a step-by-step style. Each protocol will be characterized by a brief introductory section, followed by a short aims section, in which the precise purpose of the protocol will be clarified.

Gas Chromatography-Mass Spectrometry

Based Metabolomics in Food Science

Edited by

Sastia Prama Putri

The University of Osaka, Suita, Osaka, Japan

Humana Press

Editor
Sastia Prama Putri
The University of Osaka
Suita, Osaka, Japan

ISSN 2662-950X ISSN 2662-9518 (electronic)
Methods and Protocols in Food Science
ISBN 978-1-0716-5107-0 ISBN 978-1-0716-5108-7 (eBook)
https://doi.org/10.1007/978-1-0716-5108-7

This Humana imprint is published by the registered company Springer Science+Business Media, LLC, part of Springer Nature.
The registered company address is: 1 New York Plaza, New York, NY 10004, U.S.A.

Dedication

This book is dedicated to

Professor Eiichiro Fukusaki

in sincere gratitude for his guidance and unwavering support

福﨑先生に心より感謝を込めて

Foreword

The application of metabolomics has expanded rapidly in recent years, encompassing a wide range of disciplines, including food science. In this field, metabolomics has emerged as a powerful tool for quantifying constituent compounds in food samples and elucidating their relationships to phenotypic traits. Owing to its comprehensive analytical capabilities, metabolomics is now widely employed in food science for various purposes, such as authenticating geographic origin, enhancing sensory quality, evaluating nutritional value, and ensuring food safety.

Among the analytical platforms available, gas chromatography–mass spectrometry (GC-MS) stands as a cornerstone of food metabolomics. Its relatively low operational cost, high batch-to-batch reproducibility, and broad coverage of bioactive metabolites make GC-MS particularly attractive for many applications. Nevertheless, the methodologies required for GC-MS analysis can vary significantly depending on the nature of the sample, the range of metabolites targeted, and the specific research objectives. As such, a thorough understanding of the nuances in sample preparation and method selection is critical for obtaining reliable and meaningful results. A deep understanding of sample preparation and methodological optimization is therefore essential to ensure accuracy, reproducibility, and meaningful interpretation of results.

This book is designed to serve as a comprehensive reference for researchers and professionals working in food metabolomics. It provides detailed, step-by-step guidance on GC-MS analysis procedures across the major stages of sample preparation, including extraction and derivatization techniques tailored to different types of food matrices. Each chapter focuses on specific applications and potential metabolite coverage, offering insights into alternative strategies that may be adopted to meet diverse analytical needs.

By bringing together contributions from experts with extensive experience in GC-MS-based food metabolomics, this book aims to support the advancement of knowledge and practical skills in the field. It is my sincere hope that the chapters herein will assist researchers, students, and industry professionals in refining their analytical approaches and accelerating the integration of GC-MS into innovative and impactful food science research.

Eiichiro Fukusaki

Department of Biotechnology, Graduate School of Engineering, The University of Osaka, Suita, Osaka, Japan

Industrial Biotechnology Initiative Division, Institute for Open and Transdisciplinary Research Initiatives, Suita, Osaka, Japan

The University of Osaka Shimadzu Omics Innovation Research Laboratories, International Center for Biotechnology, The University of Osaka, Suita, Osaka, Japan

Foreword

Preface to the Series

The Methods and Protocols in Food Science series is devoted to the publication of research protocols and methodologies in all fields of food science. The series is unique as it includes protocols developed, validated, and used by food and related scientists as well as provides theoretical basis for each protocol. Aspects related to improvements in the protocols and adaptations and further developments in the protocols may also be approached.

The Methods and Protocols in Food Science series aims to bring the most recent developments in research protocols in the field as well as very well-established methods. As such the series targets undergraduates, graduates, and researchers in the field of food science and correlated areas. The protocols documented in the series will be highly useful for scientific inquiries in the field of food sciences and are presented in such a way that the readers will be able to reproduce the experiments in a step-by-step style.

Each protocol will be characterized by a brief introductory section, followed by a short aims section, in which the precise purpose of the protocol is clarified. Then, an in-depth list of materials and reagents required for employing the protocol is presented, followed by comprehensive and step-by-step procedures on how to perform that experiment. The next section brings the dos and don'ts when carrying out the protocol, followed by the main pitfalls faced and how to troubleshoot them. Finally, template results will be presented and their meaning/conclusions addressed.

The Methods and Protocols in Food Science series will fill an important gap, addressing a common complaint of food scientists, regarding the difficulties in repeating experiments detailed in scientific papers. With this, the series has a potential to become a reference material in food science laboratories of research centers and universities throughout the world.

Department of Food Science and Nutrition,
Faculty of Food Engineering
University of Campinas, Campinas, Brazil

Anderson S. Sant'Ana

Preface

Gas chromatography–mass spectrometry (GC-MS) has become one of the most widely used techniques in metabolomic analysis, particularly in food science, due to its affordability, robustness, and consistent performance. Compared to other analytical platforms such as liquid chromatography–mass spectrometry (LC-MS) and nuclear magnetic resonance (NMR), GC-MS offers practical advantages, including higher reproducibility, analytical stability, and streamlined data processing.

Originally, GC-MS was primarily employed for the analysis of volatile compounds, as it is well-suited for detecting substances in the gas phase. However, significant advances in derivatization techniques have extended the capabilities of GC-MS to include the analysis of a broad range of low-molecular-weight, nonvolatile metabolites. This advancement has significantly expanded its applicability in metabolomic research.

GC-MS is particularly valued for its consistent retention times and reliable mass spectral data. Retention time reproducibility is further enhanced through the use of retention indices, calculated with reference compounds such as n-alkanes or fatty acid methyl esters, and compared against established databases including the Fiehn Library, Golm Metabolome Database, MassBank, Wiley Database, and the NIST Library. Additionally, the standardized use of electron ionization at 70 eV allows for consistent spectral matching across laboratories using shared reference libraries.

The combination of robust derivatization techniques and the inherent strengths of GC-MS as a reliable analytical platform has made it an increasingly preferred tool among researchers. As the role of GC-MS continues to expand in food metabolomics, a solid understanding of sample preparation strategies becomes critical to ensuring accurate and comprehensive metabolite profiling.

This chapter aims to provide foundational insights into sample preparation for GC-MS-based metabolomics, with a particular focus on food science applications. It serves as a practical guide to navigating the complexities of preparing diverse food matrices for effective GC-MS analysis. This volume presents standardized protocols for sample preparation in GC-MS-based metabolomics, covering 18 different types of food samples along with detailed handling instructions. The opening chapter discusses the general applications of GC-MS in food metabolomics and highlights how this approach supports robust and meaningful advances in the field.

We would like to express our sincere appreciation to all contributing authors for accepting our invitation and sharing their valuable expertise, which has been essential to the timely and successful completion of this volume. We hope that the chapters provide useful guidance on sample preparation protocols for GC-MS-based metabolomics in food science and serve as a practical reference for researchers working with similar sample types. Finally, we extend our gratitude to Springer Nature for the opportunity to contribute to this important publication.

Suita, Osaka, Japan *Sastia Prama Putri*
Suita, Osaka, Japan *Enik Nurlaili Afifah*
Suita, Osaka, Japan *Rafidha Irdiani*
R. R. C. Voluntad

Contents

Contributors

MARY FAITH YAMBALLA ADAN • *Department of Biotechnology, Graduate School of Engineering, The University of Osaka, Suita, Osaka, Japan*

ENIK NURLAILI AFIFAH • *Department of Biotechnology, Graduate School of Engineering, The University of Osaka, Suita, Osaka, Japan; Department of Agronomy, Faculty of Agriculture, Universitas Gadjah Mada, Yogyakarta, Indonesia*

WAQIF AGUSTA • *Department of Biotechnology, Graduate School of Engineering, The University of Osaka, Suita, Osaka, Japan; National Research and Innovation Agency, Jakarta, Indonesia*

FITRI AMALIA • *Research Center for Applied Microbiology, National Research and Innovation Agency, Cibinong, Bogor, West Java, Indonesia*

AULIA GUSNING ATI • *Department of Biotechnology, Graduate School of Engineering, The University of Osaka, Osaka, Japan*

TAKESHI BAMBA • *Division of Metabolomics, Medical Research Center for High Depth Omics, Medical Institute of Bioregulation, Kyushu University, Fukuoka, Japan*

GIRI ROHMAD BAROKAH • *Department of Biotechnology, Graduate School of Engineering, The University of Osaka, Suita, Osaka, Japan; Research Center for Food Technology and Processing, National Research and Innovation Agency, Gunung Kidul, Yogyakarta, Indonesia*

SIVAMOKE DISSOOK • *Department of Biochemistry, Faculty of Medicine, Chiang Mai University, Chiang Mai, Thailand*

EIICHIRO FUKUSAKI • *Department of Biotechnology, Graduate School of Engineering, The University of Osaka, Suita, Osaka, Japan; Industrial Biotechnology Initiative Division, Institute for Open and Transdisciplinary Research Initiatives, Suita, Osaka, Japan; The University of Osaka Shimadzu Omics Innovation Research Laboratories, International Center for Biotechnology, The University of Osaka, Suita, Osaka, Japan*

ABU HANIFAH • *Department of Biotechnology, Graduate School of Engineering, The University of Osaka, Osaka, Japan*

MUHAMMAD MAULANA MALIKUL IKRAM • *Department of Biotechnology, Graduate School of Engineering, The University of Osaka, Suita, Osaka, Japan*

RAFIDHA IRDIANI • *Department of Biotechnology, Graduate School of Engineering, The University of Osaka, Suita, Osaka, Japan*

ROMTEERA KITICHAIWORAKUL • *Department of Biochemistry, Faculty of Medicine, Chiang Mai University, Chiang Mai, Thailand*

RYOTA MABUCHI • *Faculty of Bioresource Sciences, Prefectural University of Hiroshima, Hiroshima, Japan*

YOSHIKA MAEKAWA • *Department of Biotechnology, Graduate School of Engineering, The University of Osaka, Suita, Osaka, Japan*

SHUNSUKE MIYAUCHI • *Department of Biotechnology, Graduate School of Engineering, The University of Osaka, Suita, Osaka, Japan*

ASUKA MORI • *Department of Biotechnology, Graduate School of Engineering, The University of Osaka, Suita, Osaka, Japan*

HIROKI NAITO • *Department of Biotechnology, Graduate School of Engineering, The University of Osaka, Suita, Osaka, Japan; The University of Osaka Shimadzu Analytical Innovation Research Laboratories, Suita, Osaka, Japan*

RAMIN PERNGMAG • *Department of Biochemistry, Faculty of Science, Chulalongkorn University, Bangkok, Thailand*

PINNAPAT PINSORN • *Center of Excellence in Molecular Crop, Department of Biochemistry, Faculty of Science, Chulalongkorn University, Bangkok, Thailand*

FARHANA R. PINU • *Biological Chemistry and Bioactives Group, The New Zealand Institute for Plant and Food Research Limited, Auckland, New Zealand*

SUPAKORN POTIJUN • *Program in Biotechnology, Faculty of Science, Chulalongkorn University, Bangkok, Thailand; Department of Biochemistry, Faculty of Science, Chulalongkorn University, Bangkok, Thailand*

SASTIA PRAMA PUTRI • *Department of Biotechnology, Graduate School of Engineering, The University of Osaka, Suita, Osaka, Japan; Industrial Biotechnology Initiative Division, Institute for Open and Transdisciplinary Research Initiatives, The University of Osaka, Suita, Osaka, Japan*

SAFIRA LATIFA ERLANGGA PUTRI • *Rohto Research Village, Rohto Pharmaceutical Co., Ltd., Kunimidai, Kizugawa, Kyoto, Japan*

DELLA RAHMAWATI • *Department of Food Technology, Faculty of Life Science and Technology, Swiss German University, Tangerang, Indonesia*

RIFQI AHMAD RIYANTO • *Department of Biotechnology, Graduate School of Engineering, The University of Osaka, Suita, Osaka, Japan; Department of Food Technology, Faculty of Agriculture, Universitas Sultan Ageng Tirtayasa, Serang, Banten, Indonesia*

MONGKON SIRIJAN • *Department of Agricultural Science, Faculty of Agriculture, Natural Resources and Environment, Naresuan University, Phitsanulok, Thailand*

SUPAART SIRIKANTARAMAS • *Center of Excellence in Molecular Crop, Department of Biochemistry, Faculty of Science, Chulalongkorn University, Bangkok, Thailand; Omics Sciences and Bioinformatics Center, Faculty of Science, Chulalongkorn University, Bangkok, Thailand*

NATHANAEL STEVEN • *Department of Biotechnology, Graduate School of Engineering, The University of Osaka, Suita, Osaka, Japan*

CHIAKI TSUSHI • *Department of Biotechnology, Graduate School of Engineering, The University of Osaka, Suita, Osaka, Japan; The University of Osaka Shimadzu Analytical Innovation Research Laboratories, Suita, Osaka, Japan*

YULIANTI • *Research Center for Food Technology and Processing, National Research and Innovation Agency, Gading, Playen, Gunung Kidul, Yogyakarta, Indonesia*

YIMENG ZHAO • *Biological Chemistry and Bioactives Group, The New Zealand Institute for Plant and Food Research Limited, Auckland, New Zealand*

Part I

General Introduction

Chapter 1

General Considerations for Method Development in GC/MS Analysis

Sastia Prama Putri, Rafidha Irdiani, Rifqi Ahmad Riyanto, and Takeshi Bamba

Abstract

Gas chromatography/mass spectrometry (GC/MS) is a powerful analytical technique that is widely used to detect and characterize low molecular weight compounds across diverse research fields, including food science. By combining the separation ability of a gas chromatograph with the identification capability of a mass spectrometer, GC/MS analysis allows the identification of a wide range of compounds. This technique offers several advantages, including fast, able to provide high resolution results, sensitive, inexpensive, and reliable when appropriate methods and components are applied. The selection of appropriate methods, instrument components, and analysis modes is critical to ensure that analysis objectives are met. Careful selection can significantly influence the accuracy and reliability of the results. Therefore, this chapter provides an overview of general considerations for method development in GC/MS analysis, highlighting the characteristics and roles of major components in analytical performance.

Key words Gas chromatography, Mass spectrometry, Materials, Methods, Parameters

1 Introduction

Chromatography is an analytical method used to separate compounds based on their interactions with a stationary phase, over which a mobile phase flows to transport compounds. In the case of gas chromatography, gas is used as the mobile phase, carrying vaporized samples. Separation in gas chromatography occurs due to different chemical characteristics of each compound, leading to specific interactions with the chromatographic column as the stationary phase. Factors influencing separation include molecular size, polarity, affinity, charge, and hydrophobicity [1]. Compounds with higher affinity for the stationary phase are retained and eluted later, while those with lower affinity pass through more quickly and reach the detector earlier. The time required for a compound to elute is referred to as its retention time [2].

Sastia Prama Putri (ed.), *Gas Chromatography-Mass Spectrometry: Based Metabolomics in Food Science*, Methods and Protocols in Food Science, https://doi.org/10.1007/978-1-0716-5108-7_1,

Gas chromatography is classified into two types—gas-liquid chromatography (GLC) and gas-solid chromatography (GSC). In GSC, the stationary phase consists of solid material, including alumina, silica, or carbon, whereas in GLC, the stationary phase is liquid adsorbed onto a solid inert packing on the capillary tubing walls [3, 4]. In GSC, polar molecules tend to interact strongly with the solid stationary phase, resulting in prolonged retention and severe peak tailing in the chromatograms [5]. These limitations have led to a more widespread use of GLC across most fields of science. At present, the term "GC" is generally used to refer to GLC [6].

The mass spectrometer (MS) is an instrument used to detect and identify compounds within a sample mixture. When coupled with gas chromatography (GC), the MS consists of several key components, including a vacuum pump, sample introduction unit, ionization chamber, mass analyzer, and detector. In the sample injection unit, compounds that have been separated by GC enter the MS. These compounds are then ionized in the ionization chamber. For GC/MS analysis, the most commonly employed ionization method is electron ionization (EI), also referred to as hard ionization. In EI, a heated filament generates electrons, which bombard the incoming molecules, producing molecular ions. This process typically causes fragmentation, in which molecules are broken down into a fragment ion and a neutral fragment. The resulting fragmentation patterns are highly reproducible and unique to each compound, serving as a "fingerprint" for annotation. After ionization, the fragment ions go into the mass analyzer and are subsequently detected by the detector [7].

When combined, gas chromatography/mass spectrometry (GC/MS) is a powerful analytical technique that is widely used to analyze biological samples. The combination of gas chromatography and mass spectrometry enables both qualitative and quantitative evaluations of the chemical constituents in a sample. In GC/MS, compounds are first separated by the gas chromatograph, while the mass spectrometer detects and characterizes each compound based on its mass-to-charge ratios [8].

GC/MS offers several advantages, including fast, able to provide high resolution results, sensitive, inexpensive, and reliable [7]. However, the quality and the reliability of the data are dependent on parameter selection. Careful selection and consideration are necessary to improve analytical performance. Therefore, in this chapter, the principles and key components of GC/MS will be discussed briefly, aiming to provide readers with fundamental information and general considerations in conducting GC/MS analysis.

2 Materials Selection

GC-MS instruments consist of several components that work together to detect and characterize compounds. Key elements, including carrier gas, columns, and mass analyzers, offer multiple options that can be applied depending on the analysis objectives and practical considerations. This section discusses the characteristics and roles of these components.

2.1 Carrier Gas Selection

In gas chromatography, the mobile phase is also known as *carrier gas*. Carrier gas must have a high purity in order to provide good chromatograms with minimum baseline noise and to maintain column performance. In addition, the gas used must meet several criteria, such as being inert, having no interaction with the target compounds and the stationary phase, being compatible with the detection method, and being able to diffuse quickly across the column to promote high separation [2, 9].

Several gases are commonly used as carrier gas, including helium, hydrogen, nitrogen, and argon. Of these, helium and hydrogen are the most widely used, with helium being the predominant choice in many laboratories. Some advantages and disadvantages of each carrier gas are summarized in Table 1.

Other than the type of carrier gas, several parameters including temperature and carrier gas flow are also necessary to be

Table 1
Advantages and disadvantages of each carrier gas

Gas	Advantages	Disadvantages	References
Helium	Easy to handle Nonflammable Wide flow range Compatible with almost all types of detectors	High cost Limited and nonrenewable	[10]
Hydrogen	Great separation efficiency Faster analysis Improved resolution Relatively affordable Stable supply	Flammable Rapid expansion at high pressure Corrosive potential to the ion source More background noise	[11]
Nitrogen	Stable supply Higher separation efficiency than helium and hydrogen at its optimum linear velocity Greater vacuum pump efficiency Relatively affordable	Low sensitivity Longer analysis time is necessary	[10, 12]
Argon	Greater vacuum pump efficiency Less expensive than helium	More expensive than nitrogen	[13]

considered. Temperature affects the ratio of the viscosity to diffusion coefficient. The diffusion coefficient describes how quickly molecules (the carrier gas or analytes within it) spread out due to random molecular motion inside the column. Low viscosity combined with a high diffusion coefficient is preferred as this promotes high separation efficiency [14].

Meanwhile, the carrier gas flow is important for column efficiency. A stable and reproducible flow rate is essential so that the retention times can be reproduced for reliable compound identification [7]. The carrier gas flow rate (f) can be calculated using Eq. 1, with r represents the radius of the column (mm) and u describes the linear velocity of the carrier gas in cm/s. The linear velocity can be determined by dividing the column length (cm) by the time (s) required for methane, butane, or other compound that is not retained by the column to travel from the point of injection to the detector [9].

$$f = 0.6u\pi r^2 \tag{1}$$

2.2 Column Selection

The column is a critical component of a gas chromatograph, as it plays the main role in separation. There are two types of columns that are mainly used—packed columns and capillary columns.

Packed Columns

Packed columns are the first generation of GC columns. They have an outside diameter of 0.32 or 0.64 cm and are compatible with both GSC and GLC. They can be made of glass, stainless steel, nickel, or teflon. The main characteristic of packed columns is the presence of fillers. The fillers contain some solid adsorbent coated with a solid or liquid stationary phase [3].

Capillary Columns

At present, most GC instruments are compatible with capillary columns. These columns usually have an inner diameter of 0.1–0.5 mm. Unlike packed columns, capillary columns are only compatible with GLC and do not have any fillers. Instead, the inner surface of the column is coated with a thin film of liquid stationary phase. Capillary columns can be divided into three types: porous layer open tubular (PLOT) columns, wall-coated open tubular (WCOT) columns, and support-coated open tubular columns (SCOT). In PLOT columns, the stationary phase is coated on a porous layer of a solid porous layer, while in WCOT, it is directly coated onto the column wall. As for SCOT, the stationary phase is supported by an adsorbed layer of solid particles [3].

The choice of columns depends on the analysis goal. Capillary columns are mainly used for trace analysis in complex mixtures, whereas packed columns are usually used for analysis of gases, volatile organic compounds (VOCs), and compounds with high molecular weight [15]. Furthermore, the column selection should also consider the polarity of the stationary phase, column length, and column diameter [2]. Longer columns generally provide higher resolution but slower analysis duration. Meanwhile, the column diameter affects the resolution, speed, and capacity of the columns. A smaller diameter typically results in better resolution and faster analysis [3].

2.3 Types of Mass Analyzers

In a mass analyzer, fragment ions generated in the ionization chamber are separated based on their mass-to-charge ratio (m/z). There are several types of mass analyzers. Quadrupole and time-of-flight (TOF) analyzers are the most widely used.

Quadrupole

A quadrupole contains four parallel cylindrical metal rods (electrodes with a hyperboloidal interior surface) inside a vacuum chamber, positioned equidistant from the center axis. Both direct current (DC) and high frequency alternating current or radiofrequency (RF) are applied to the quadrupole, so that only the ions with the target m/z successfully pass through the quadrupole and get to the detector. When a given set of parameters is applied to the poles, certain ions of a specific m/z range maintain a stable oscillation and pass through the quadrupole to reach the detector (resonance ion). The oscillations of ions with other m/z values become unstable, causing them to collide with the poles, go outside of the system, and not be detected (non-resonance ion). The key features of a quadrupole are small and relatively cheap. However, accurate mass cannot be obtained due to low resolution [2].

Time-of-Flight

As for the TOF analyzer, it works based on the measurement of the time needed for the ions to reach the detector from the ionization chamber. The difference in the time needed is due to different m/z results in different velocities. The key features of this analyzer are high sensitivity and better resolution compared to a quadrupole. Despite these advantages, TOF instruments do have certain limitations. Their relatively large size and higher cost make them less practical for routine laboratory use, especially when compared to quadrupole analyzers, which are more compact, cost-effective, and commonly employed in standard GC-MS workflows [2].

3 Method Selection

Similar to material selection, selection of appropriate derivatization methods and injection configurations are necessary depending on the target metabolites and sample characteristics. This section outlines their principles and key features.

3.1 Derivatization Methods

The development of derivatization method allows GC to be able to analyze nonvolatile compounds. Derivatization is defined as a chemical reaction that aims to produce a more volatile derivative by modifying the compound with the help of derivatization reagents [16]. There are several criteria that need to be fulfilled by a derivatization reagent, including being able to generate more than 95% complete derivatives, not resulting in structural rearrangement, not causing loss of analyte during reaction, having a side product of the reaction that will not interact with the GC column, and producing a derivative that is stable [17]. There are several derivatization methods that can be applied depending on the target compounds. The summary is included in Table 2 [2, 17, 18].

3.2 Injection Configuration

In GC, the sample is introduced to the column through the injection port. An important factor to consider during sample injection is the injection configuration, which can be performed in either split or splitless mode.

Table 2
Derivatization method for GC/MS analysis

No	Method	Target group	Reagent	Group to be added
1	Oximation	-CO-	Methoxyamine hydrochloride	-C=N-OR
2	Silylation	-OH, -SH $-NH_2$, -NH -COOH	N-Methyl-N-(trimethylsilyl)trifluoroacetamide (MSTFA), O-bis (trimethylsilyl)-trifluoroacetamide (BSTFA), etc.	$Si(CH_3)_3$, etc.
3	Acylation	-OH, -SH $-NH_2$, -NH -C=C-	Trifluoroacetic anhydride (TFA), pentafluoropropionic anhydride (PFP), heptafluorobutyric anhydride (HFB), etc.	R–C=O, etc.
4	Esterification	-COOH	Methanol/HCl, diazomethane (CH_2N_2), pentafluorobenzyl bromide (PFBBr), etc.	-COOR, etc.

Split Mode

In split mode, only a portion of the sample is injected into the column. The rest of the sample is mixed with the carrier gas and is purged through the waste port. This mode enables the prevention of column overload and allows for quick sample injection.

Splitless Mode

In splitless mode, the entire sample is introduced to the column. This mode is useful when analyte concentrations are too low to be effectively detected by using split mode [2, 7].

4 Notes

Ultimately, there is no universal GC/MS analysis method that can be applied to all types of analyses. Instead, reliable results can be obtained through careful analysis design by considering sample characteristics, target metabolites, and instrument capabilities. By understanding the principles, characteristics, and roles of each component, researchers can not only design GC/MS methods that align with the analysis objectives but also expand the scope of applications in the field of food science.

References

1. Kumari VBC, Patil SM, Ramu R, Shirahatti PS, Kumar N, Sowmya BP, Egbuna C, Uche CZ, Patrick-Iwuanyanwu KC (2022) Chromatographic techniques: types, principles, and applications. In: Analytical techniques in biosciences. Elsevier, London pp 73–101
2. Sánchez-Guijo A, Hartmann MF, Wudy SA (2013) Introduction to gas chromatography-mass spectrometry. In: Hormone assays in biological fluids. Springer, New Jersey pp 27–44
3. Rahman MM, Abd El-Aty AM, Choi J, Shin H, Shin SC, Shim J (2015) Basic overview on gas chromatography columns. In: Analytical separation science, pp 823–834. https://doi.org/10.1002/9783527678129.assep024
4. Seneca (2007) Alkaloid chemistry. In: Alkaloids – secrets of life, pp 61–139. https://doi.org/10.1016/B978-044452736-3/50004-0
5. Falaki F (2019) Sample preparation techniques for gas chromatography. In: Gas chromatography-derivatization, sample preparation, application, pp 1–30. https://doi.org/10.5772/intechopen.84259
6. Shellie RA, Heng WS (2023) Gas chromatography. In: Encyclopedia of forensic sciences, vol 1–4, 3rd edn, pp 1–9. https://doi.org/10.1016/B978-0-12-823677-2.00234-8
7. McNair HM, Miller JM, Snow NH (2019) Basic gas chromatography. John Wiley & Sons, USA
8. Medeiros PM (2018) Gas chromatography–mass spectrometry (GC–MS). In: Encyclopedia of geochemistry. Springer, Switzerland pp 530–535
9. Sparkman OD, Penton Z, Kitson FG (2011) Gas chromatography and mass spectrometry: a practical guide. Academic Press, Cambridge
10. Oshima N, Takagi M, Sakai S, Ikarashi Y (2022) Comparison of helium-alternative carrier gases for gas chromatography/mass spectrometry of standard test methods for indoor air quality guidelines in Japan. BPB Rep 5(4): 84–87
11. Çetintürk K, Güzel B, Canlı O (2025) The development of a novel, green, efficient, and eco-friendly GC-MS/MS analytical method for

the reliable and rapid determination of dl-PCBs, and PCDD/Fs using hydrogen as a carrier gas and a modified ion source. Talanta 283:127180. https://doi.org/10.1016/J.TALANTA.2024.127180

12. Saito-Shida S, Saito M, Tsutsumi T (2024) Comparison of nitrogen and helium as carrier gases for determination of pesticide residues in foods via gas chromatography–tandem mass spectrometry with atmospheric pressure chemical ionization. J Food Compos Anal 134: 106492. https://doi.org/10.1016/J.JFCA.2024.106492
13. Paul S, Saikia A, Majhi V, Pandey VK (2021) Analytical instruments. In: Introduction to biomedical instrumentation and its applications, pp 213–250. https://doi.org/10.1016/B978-0-12-821674-3.00009-7
14. Grob RL, Barry EF (2004) Modern practice of gas chromatography. John Wiley & Sons, Hoboken
15. Akash MSH, Rehman K (2025) Comprehensive insights into gas chromatography. In: Essentials of pharmaceutical analysis. Springer, pp 757–796. https://doi.org/10.1007/978-981-96-5996-8_15
16. Jan S, Ahmad P (2019) Ecometabolomics: metabolic fluxes versus environmental stoichiometry. Academic Press. https://doi.org/10.1016/C2017-0-01253-3
17. Orata F (2012) Derivatization reactions and reagents for gas chromatography analysis. In: Advanced gas chromatography-progress in agricultural, biomedical and industrial applications, IntechOpen. p 91. https://doi.org/10.5772/33098
18. Moldoveanu SC, David V (2018) Derivatization methods in GC and GC/MS. In: Gas chromatography-derivatization, sample preparation, application. IntechOpen, London

Part II

Fruits and Agricultural Products

Chapter 2

GC-MS-Based Metabolite Profiling Method for Chaya Leaf

Supakorn Potijun, Ramin Perngmag, and Supaart Sirikantaramas

Abstract

The analysis of plant components, particularly those from leafy vegetables used for food or pharmacological purposes, has gained significant attention due to their potential applications. Attributes such as color and flavor are often studied to optimize these resources, especially amid growing food security concerns. Metabolite profiling, which focuses on primary metabolites like sugars, amino acids, and organic acids, has become increasingly important, as these metabolites are directly or indirectly linked to flavor. This chapter outlines methods for metabolite profiling of *Cnidoscolus* spp. (chaya), plants known for their distinct umami flavor derived from amino acids, peptides, and nucleotides. Gas chromatography–mass spectrometry plays a pivotal role in this analysis, enabling the examination of volatile and nonvolatile compounds and providing valuable insights into their metabolite compositions relevant to food science research.

Key words Chaya, Gas chromatography–mass spectrometry, Metabolites, Vegetable, Volatiles

1 Introduction

Chaya (*Cnidoscolus chayamansa* and *C. aconitifolius*), also known as Mayan spinach tree (Fig. 1), is highly regarded for its culinary and medicinal value, particularly in Central America and Mexico, where it is commonly used in soups, stews, and beverages. Known for its umami flavor—attributed to its amino acid and peptide contents—chaya has earned the title of "superfood" due to its numerous health benefits, including lowering blood sugar and fat levels and regulating blood pressure [1]. However, due to its cyanide content, chaya must be cooked before consumption to ensure safety [2].

Given its rich nutritional and pharmacological properties, chaya is an ideal candidate for metabolite analysis using gas chromatography–mass spectrometry (GC–MS). A previous GC–MS study provided a comprehensive metabolic profile of chaya leaves, revealing that young leaves exhibit an intense umami taste, with high levels of L-glutamic acid, 5′-adenosine monophosphate, and novel small peptides contributing to the flavor [2]. Another

Sastia Prama Putri (ed.), *Gas Chromatography-Mass Spectrometry: Based Metabolomics in Food Science*, Methods and Protocols in Food Science, https://doi.org/10.1007/978-1-0716-5108-7_2,

Fig. 1 Chaya leaf (**a**), *C. chayamansa* (**b**) *C. aconitifolius*

report comparing dried and fresh chaya leaves identified 3-methylbutanal, a malt-like aroma compound, as a key volatile in dried leaves. Additionally, the drying process increased the 5-′-ribonucleotide content threefold [3].

These findings highlight the utility of GC–MS in analyzing volatile and nonvolatile compounds, offering insights into chaya's biochemical composition and metabolic pathways, including amino acid synthesis, carbohydrate metabolism, and organic acid pathways. GC–MS also identifies volatile compounds such as terpenes, alcohols, and aldehydes [4, 5], which shape chaya's flavor profile, making it valuable for both culinary and pharmacological applications. Proper sample preparation is crucial for metabolite analysis. Freeze-drying is preferred for metabolomics to preserve heat-sensitive compounds, whereas heat-drying is suitable for studying heat-induced aroma changes. This methodology enhances our understanding of chaya's metabolites and serves as a versatile tool for studying other plants [6–8].

2 Materials

2.1 Plant Sample

Chaya leaf (dried with various methods, such as freeze-drying, vacuum drying, oven drying, or pan roasting).

2.2 Chemicals

1. D-(-)-13C-sorbitol (internal standard).
2. 3-Methyl-2-heptanone (internal standard for headspace).
3. MSTFA (N-Methyl-N-(trimethylsilyl)trifluoroacetamide), BSTFA (N,O-Bis(trimethylsilyl)trifluoroacetamide) (derivatizing agent).

4. Methanol (LC-grade).
5. Water (LC-grade).
6. Chloroform (LC-grade).
7. Methoxyamine hydrochloride.
8. Pyridine.
9. Alkane standard (C10–C40 or C5–C12 based on molecular weight of focused metabolites).
10. Dry nitrogen gas (for sample drying) and liquid nitrogen.

2.3 Equipment

1. Vials and insert.
2. Pipettes and tips.
3. Analytical balance (accuracy ±0.1 mg).
4. Centrifuge.
5. Vortex mixer.
6. Thermomixer.
7. Speed vacuum or CentriVap.
8. GC-MS system with column.
9. Miscellaneous.
10. Filter paper or 0.45 μm filter.
11. Syringe and needle.
12. Procedure.
13. Vials and insert.

3 Methods

Figure 2 shows the flowchart for nonvolatile and volatile compound analyses of chaya leaves.

3.1 Protocols: Nonvolatile Metabolome Analysis of the Chaya Leaf Samples

This protocol outlines the procedure for analyzing nonvolatile metabolites in chaya leaf samples using GC-MS, focusing on compounds such as sugars, amino acids, organic acids, and peptides, which are nonvolatile but essential for metabolic profiling [6–8].

3.1.1 Sample Preparation

1. Collect fresh plant leaf samples or use freeze-dried material.
2. Homogenize the plant leaves using a mortar and pestle with liquid nitrogen or a blender for fresh samples. For freeze-dried samples, grind the material using a mechanical grinder (mixer mill) at 30 Hz for 45 s.
3. Weigh approximately 25 mg of leaf tissue and transfer it into a 2 mL microcentrifuge tube.

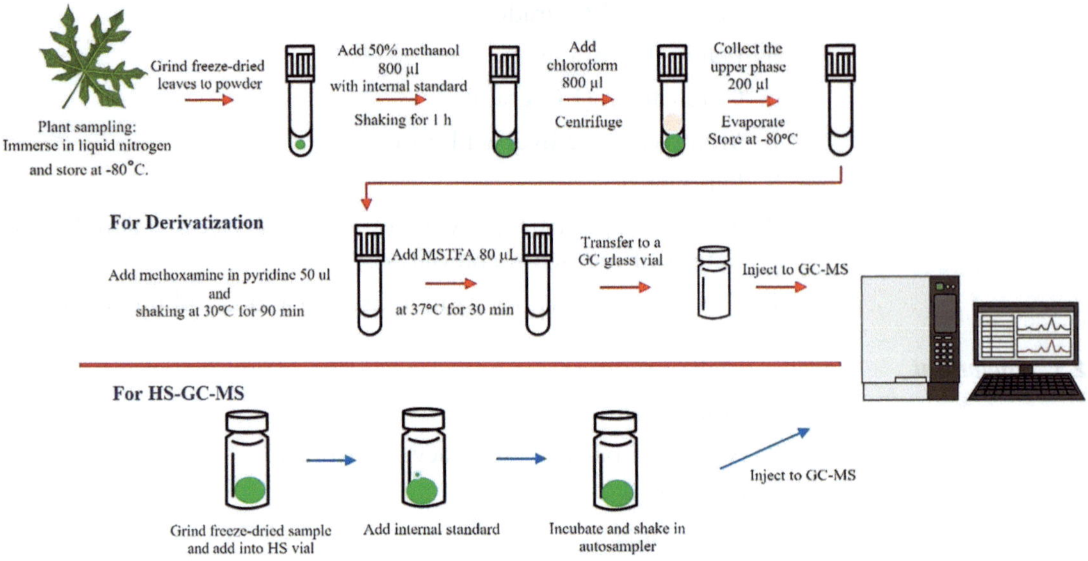

Fig. 2 Flowchart for GC-MS and HS-GC-MS analyses of chaya leaves

3.1.2 Extraction of the Metabolites

1. Add 800 μL of 50% methanol (v/v), precooled to −20 °C, containing D-(-)-13C-sorbitol (0.015 mg/mL).
2. Vortex for 30 s.
3. Incubate at 25 °C for 1 h at 500 rpm.
4. Centrifuge at 8000 rpm for 5 min.
5. Filter through a 0.45-μm filter.
6. Add 800 μL of chloroform, vortex, and centrifuge.
7. Collect 200 μL of the upper phase, transfer it to a new tube, and concentrate using a SpeedVac at 25 °C.
8. Blow nitrogen gas over the sample, then store at −20 °C.

3.1.3 Derivatization

1. Dissolve methoxyamine hydrochloride in pyridine (20 mg/mL), then add 50 μL to the sample. Incubate at 30 °C for 90 min.
2. Add 80 μL of BSTFA or MSTFA (preferably MSTFA) and incubate at 37 °C for 30 min.
3. Allow the sample to cool before injection.

3.1.4 GC-MS Analysis

1. GC-MS with an appropriate column (e.g., DB5-MS, 30 m × 0.25 mm, 0.25-μm film thickness) (Table 1).
2. Inject 1 μL of the derivatized sample.
3. Program the GC oven temperature and MS settings according to Table 2.

Table 1
GC-MS column

Column	Stationary phase	Length	Diameter (mm)	Film thickness (μm)	Max temp. (°C)
DB5-MS (Agilent)	5% phenyl, 95% dimethylpolysiloxane	30 m (typical)	0.25	0.25	350
RTX-5MS (Restek)	5% phenyl, 95% methyl polysiloxane	30 m (typical)	0.25	0.25	325
TG-5MS (Thermo)	5% phenyl, 95% methyl polysiloxane	30 m (typical)	0.25	0.25	325

Table 2
GC-MS setup for plant leaf metabolite analysis

Parameter	Value/description
GC system	Agilent 7890A, Shimadzu QP2020 or equivalent
Column type	DB5-MS, RTX-5MS, or TG-5MS (30 m × 0.25 mm, 0.25-μm film thickness) or equivalent
Carrier gas	Helium (99.999% pure), constant flow mode
Flow rate	1.2 mL/min
Injection volume	1 μL
Injection mode	Splitless and split (20:1)
Injection temperature	250 °C
Detector	Mass spectrometer (MS)
Ionization energy	70 eV
MS scan range	50–400 m/z
Transfer line temperature	280 °C
GC cycle time	45 min
MS source temperature	230 °C
Mass spectrometer mode	Full scan mode or SIM (selected Ion monitoring) depending on the targeted metabolites
Solvent delay	3 min (to avoid solvent peak interference)
Carrier gas pressure	10 psi (or optimized for the column and system)

3.1.5 Data Analysis

1. Annotate metabolites by comparing their mass spectra and retention times to known standards or reference databases, such as the NIST or Wiley library. A similarity match of >80% is typically required for accurate identification.
2. Perform statistical analysis to interpret and compare differences in metabolite profiles between different plant samples or experimental conditions.

3.1.6 Quality Control and Replication

1. Include blanks (methanol) and internal standards in each sample batch to ensure accuracy and reproducibility.
2. Perform at least three biological and three technical replicates for each plant sample.
3. Calculate the retention indices (RIs) for each metabolite using an alkane standard to ensure the accuracy and consistency of the retention times.
4. Quantify the identified metabolites using internal standards for normalization to account for variations in sample preparation and injection volume.

3.2 Protocols: Volatile Metabolome Analysis of the Chaya Leaf Samples

This protocol outlines the procedure for analyzing volatile metabolites, such as terpenes, alcohols, aldehydes, and other aromatic compounds, which are responsible for the plant's aroma [4, 5].

3.2.1 Sample Preparation

1. For freeze-dried samples, grind the material using a mechanical grinder (mixer mill) at 30 Hz for 45 s, and weigh approximately 0.3 g for volatile analysis.
2. For fresh samples, collect fresh leaves and weigh approximately 3–5 g for volatile analysis. Cut or gently crush the leaves to release volatiles.
3. Transfer the material into a 20 mL headspace vial.
4. Add 5 μL of 3-methyl-2-heptanone (1000 ppm) to each sample as an internal standard.

3.2.2 Headspace Sampling

1. Seal the vial tightly with a septum cap.
2. Equilibrate the vial at 60 °C for 30 min to allow the volatiles to partition into the headspace.
3. Place the vial into the GC-MS autosampler tray.
4. Sample 500 μL of the headspace into the injector of the GC-MS system.

3.2.3 GC-MS Analysis

1. Program the GC to use a DB-5MS column or an equivalent (30 m × 0.25 mm × 0.25 μm).

2. Program the GC oven temperature and MS according to Table 3.

3.2.4 Data Analysis

Analyze the MS spectra to annotate compounds by comparing their mass-to-charge (m/z) ratios with reference spectra from standards or reference databases, such as the NIST or Wiley library. A match of >80% similarity is typically required for accurate identification.

Table 3
HS-GC-MS setup for plant leaf metabolite analysis

Parameter	Value/description
GC system	Agilent 7890A, Shimadzu QP2020, or equivalent with headspace
Column type	DB5-MS, RTX-5MS, or TG-5MS (30 m × 0.25 mm, 0.25-μm film thickness) or equivalent
Carrier gas	Helium (99.999% pure), constant flow mode
Trap desorption temperature	220 °C
Trap oven temperature	80 °C
Equilibration time	15 min
Multi-injection count	10
Pressurizing gas pressure	80 kPa
Load time	0.5 min
Load equilibrium time	0.1 min
Injection time	0.5 min
GC cycle time	40 min
Column flow (He)	1 mL/min
Injection mode	Splitless
Column oven temperature program	Start at 35 °C (hold 4 min), then increase to 180 °C at 15 °C/min, then increase to 250 °C at 40 °C/min (hold 5 min)
Ion source temperature	230 °C
Interface temperature	250 °C
Detector gain	0.87 kV
Event time	0.3 s
Scan acquisition mode	SCAN
Scan speed	1428 μ/s
Mass scan range (m/z)	29–400

3.2.5 Quality Control and Replication

1. Include blanks (ultrapure water) and internal standards in each sample batch to ensure accuracy and reproducibility.
2. Perform at least three biological and three technical replicates for each plant sample.
3. Calculate the retention indices (RIs) for each metabolite using an alkane standard (for headspace analysis), use low-molecular-weight alkanes (C5–C12) to ensure the accuracy and consistency of retention times.
4. Quantify the annotated metabolites using internal standards for normalization to account for variations in sample preparation and injection volume.

4 Notes

Below are cautions and recommendations for experimentation.

4.1 Tips for Accuracy

1. Sample Collection

   Upon collection of plant samples, they should be promptly immersed in liquid nitrogen and subsequently stored at −80 °C to halt all metabolic processes within the cells, preserving the integrity of the compounds and preventing their degradation.
2. Weighing and Measuring the Substances

   Ensure the highest accuracy using a high-precision balance and high-quality pipettes to obtain reliable results.
3. Preparation of the Extraction Solutions

   Always prepare fresh solutions to ensure that the 50% methanol concentration is correct. Add the internal standard as the last step before use.

4.2 Tips for Handling Samples

1. Handling of Polar-Phase Solutions

   Avoid drawing up any nonpolar phase to prevent contamination.
2. Oxidation Prevention

   After extraction and complete evaporation of the solvent, use nitrogen gas blowing to prevent oxidation before the analysis.
3. Effects of Humidity on Derivatization

   Humidity can affect the efficiency of derivatization. If possible, perform the process in a controlled-humidity environment or avoid conducting it during periods of high relative humidity.
4. Injection of Samples After Derivatization

Inject the sample into the GC-MS system within 24 h of derivatization to ensure accuracy.

5. Preventing Evaporation in HS-GC-MS

   When placing the sample in an HS vial, seal the lid tightly and quickly to prevent the evaporation of the volatile compounds.

4.3 Tips for Safety

Chemicals used in the derivatization process can be harmful to the skin and respiratory system. Always wear gloves and perform the work in a chemical fume hood to ensure safety.

4.4 Tips for Maintaining the Equipment

Check the GC-MS system before each use. Perform auto-tune regularly, clean the syringe when switching sample groups, and replace the septum as scheduled to maintain optimal instrument performance.

4.5 Tips for Experiment Planning (in Case of Nonvolatile Compounds)

1. 2-Day Before: Prepare samples in advance to ensure they are ready for extraction.
2. 1-Day Before: Perform extraction and use SpeedVac for evaporation overnight to concentrate the samples.
3. D-Day

   Morning: Conduct derivatization to prepare samples for analysis.

   Afternoon: Inject the samples into the GC-MS system for analysis (overnight, if needed)

   Next Day: Collect the results and interpret the data.

References

1. Guevara-Cruz M, Medina-Vera I, Cu-Cañetas TE, Cordero-Chan Y, Torres N, Tovar AR, Márquez-Mota C, Talamantes-Gómez JM, Pérez-Monter C, Lugo R, Gutiérrez-Solis AL, Avila-Nava A (2021) Chaya leaf decreased triglycerides and improved oxidative stress in subjects with dyslipidemia. Front Nutr 8:666243. https://doi.org/10.3389/fnut.2021.666243
2. Hutasingh N, Tubtimrattana A, Pongpamorn P, Pewlong P, Paemanee A, Tansrisawad N, Siripatrawan U, Sirikantaramas S (2023) Metabolite profiling and identification of novel umami compounds in the chaya leaves of two species using multiplatform metabolomics. Food Chem 404:134564. https://doi.org/10.1016/j.foodchem.2022.134564
3. Hutasingh N, Tubtimrattana A, Pongpamorn P, Pewlong P, Paemanee A, Tansrisawad N, Siripatrawan U, Sirikantaramas S (2024) Unraveling the effects of drying techniques on chaya leaves: metabolomics analysis of nonvolatile and volatile metabolites, umami taste, and antioxidant capacity. Food Chem 446:138769. https://doi.org/10.1016/j.foodchem.2024.138769
4. Deng H, He R, Huang R, Pang C, Ma Y, Xia H, Liang D, Liao L, Xiong B, Wang X, Zhang M, Ao X, Yu B, Han D, Wang Z (2022) Optimization of a static headspace GC-MS method and its application in metabolic fingerprinting of the leaf volatiles of 42 citrus cultivars. Front Plant Sci 13:1050289. https://doi.org/10.3389/fpls.2022.1050289
5. Silva ÉAS, Saboia G, Jorge NC, Hoffmann C, dos Santos Isaias RM, Soares GL, Zini CA (2017) Development of a HS-SPME-GC/MS protocol assisted by chemometric tools to study herbivore-induced volatiles in Myrcia splendens.

Talanta 175:9–20. https://doi.org/10.1016/j.talanta.2017.06.063
6. Waris M, Kocak E, Gonulalan EM, Demirezer LO, Kır S, Nemutlu E (2022) Metabolomics analysis insight into medicinal plant science. Trends Anal Chem 157:116795. https://doi.org/10.1016/j.trac.2022.116795
7. Lisec J, Schauer N, Kopka J, Willmitzer L, Fernie AR (2006) Gas chromatography mass spectrometry–based metabolite profiling in plants. Nat Protoc 1:387–396. https://doi.org/10.1038/nprot.2006.59
8. Fiehn O (2016) Metabolomics by gas chromatography-mass spectrometry: combined targeted and untargeted profiling. Curr Protoc Mol Biol 114:30.4.1–30.4.32. https://doi.org/10.1002/0471142727.mb3004s114

Chapter 3

GC-MS-Based Metabolite Profiling Method for Cocoa

Enik Nurlaili Afifah, Aulia Gusning Ati, and Abu Hanifah

Abstract

The quality of the cacao bean is determined by its biochemical compounds, including both related to nonvolatile and volatile components. This makes the study of metabolomic profiles in cacao an essential approach to identifying key metabolites associated with high-quality cocoa products. Metabolomics is a powerful and robust method for uncovering biochemical compounds present in cacao bean samples. However, the metabolites detected can vary significantly depending on the analytical method used. This chapter provides a comprehensive overview of metabolomic analysis for cacao beans, focusing on both volatile and nonvolatile compounds. Gas chromatography-mass spectrometry (GC-MS) is highlighted as a versatile analytical tool that can be applied to both volatile compounds and nonvolatile compounds through derivatization. The chapter outlines essential procedures, including sample preparation, extraction techniques, and analytical conditions, while emphasizing critical factors for optimizing the analysis to ensure reliable and meaningful results.

Key words Analytical method, Cocoa, Metabolites, Nonvolatile, Volatile

1 Introduction

Cacao beans, the seeds of *Theobroma cacao*, are among the most significant agricultural commodities globally, primarily serving as the key ingredient in chocolate production [1]. Cocoa products are highly valued for their delicate taste and versatility as food ingredients, contributing to their widespread consumption [2]. Consequently, the quality of cacao beans is a critical factor in the market, attracting considerable attention from researchers aiming to understand and enhance cocoa quality [3].

Research on cacao spans various objectives, with a major focus on quality parameters [3]. These studies investigate the influence of environmental factors [4], postharvest processes [5, 6], cacao genotypes [7], and antioxidant properties on cocoa quality [8]. Notably, the chemical composition of the cacao bean plays a fundamental role in determining its quality, making it a popular subject for metabolomic studies.

Sastia Prama Putri (ed.), *Gas Chromatography-Mass Spectrometry: Based Metabolomics in Food Science*, Methods and Protocols in Food Science, https://doi.org/10.1007/978-1-0716-5108-7_3,

In recent years, metabolomic studies on cocoa have increased significantly, aiming to identify quality markers in cocoa [9, 10]. For instance, studies have examined the difference between fine-flavored cocoa and bulk or standard-quality cocoa [11–13], the effects of fermentation processes [5], pulp reduction, and pre-drying treatments on improving cocoa quality [14]. Metabolomics has been a central tool in these investigations, employing various analytical techniques, such as gas chromatography-mass spectrometry (GC-MS), liquid chromatography-mass spectrometry (LC-MS), and nuclear magnetic resonance (NMR) [8].

Among these techniques, GC-MS-based metabolomics is particularly notable for its high sensitivity in detecting both volatile and non-volatile metabolites [15], which are key to assessing flavor quality and monitoring metabolite changes during processing [16]. This method is widely used for analyzing agricultural products, including cocoa [4] and coffee [17, 18], due to its reproducibility, stability, and relatively straightforward data analysis [19].

Metabolomic studies on cocoa often report varying numbers of detected metabolites [20]. This variability is frequently attributed to differences in analytical instruments, sample preparation, extraction methods, and derivatization protocols [8]. Unfortunately, many studies lack detailed descriptions of these methodologies, limiting reproducibility and the ability to detect a comprehensive range of metabolites. Thus, providing standardized protocols for the metabolomic analysis of cocoa, including detailed steps for sample extraction, derivatization, and instrument settings, would be highly beneficial. Such protocols can enhance the reliability of studies and improve the detection of metabolites of interest. This chapter aims to address this gap by presenting detailed procedures for GC-MS-based metabolomics of cacao beans, enabling researchers to achieve more robust and comparable results.

2 Materials

2.1 Sample

1. Cocoa bean (raw and/or fermented).
2. Roasted cocoa beans or cocoa nibs.
3. Cocoa liquor or paste.
4. Cocoa cake (result of pressed cocoa liquor).
5. Chocolate bar.
6. Cocoa powder (cocoa drink).

2.2 Nonvolatile Analysis

2.2.1 Reagents

1. Liquid nitrogen.
2. Methanol (LC-MS grade).
3. Chloroform (LC-MS grade).

4. Ultrapure water.
5. Ribitol (as internal standard).
6. Methoxyamine hydrochloride (98+%).
7. Ultrapure pyridine (99.5+%).
8. N-methyl-N-(trimethylsilyl)-trifluoroacetamide (MSTFA).
9. C8–C40 alkane standard mixture.

2.2.2 Laboratory Supplies

1. Multi-beads shocker and 50 mL polycarbonate tube.
2. Micropipette.
3. Freezer-safe marker pen for labeling the samples.
4. Freezer gloves for handling frozen objects.
5. −30 °C freezer to store samples.
6. 2.0 mL microcentrifuge tubes.
7. 1.5 mL microcentrifuge tubes.
8. 15 mL centrifuge tubes (for high-volume samples).
9. Vortex mixer.
10. Refrigerated centrifuge.
11. Thermomixer.
12. Vacuum concentrator.
13. GC-MS vials equipped with 200 μL insert.

2.2.3 GC-MS Equipment

1. GC-MS system: GCMS-QP2010 Ultra (Shimadzu) or equivalent system.
2. GC capillary column: InertCap 5 MS/NP column (GL Sciences) or equivalent column.
3. NIST library (NIST/EPA/NIH EI-MS Library).
4. GL-Science DB (InertCap 5MS Metabolomics, Kovats RI)—can be downloaded on the MS-DIAL website.

2.3 Volatile Analysis

2.3.1 Reagents

1. Liquid nitrogen.
2. Fatty acid ethyl esters (FAEEs, C4–C24).
3. Hexane 98%.

2.3.2 Laboratory Supplies

1. Multi-bead shocker.
2. 50 mL polycarbonate tube.
3. Micropipette.
4. Freezer-safe marker pen for labeling the samples.
5. Freezer gloves for handling frozen objects.
6. −30 °C Freezer to store samples.

7. GC-MS vials equipped with 200 μL insert.
8. A SPME vial (20 mL screw-cap vial).

2.3.3 GC-MS Equipment

1. GC-MS system: GCMS-TQ8050 NX (Shimadzu) or equivalent system.
2. GC capillary column: InertCap FFAP, 0.32 mm × 60 m, 0.50 μm (GL Sciences).
3. A DVB/CAR/PDMS.
(Divinylbenzene/Carboxen/Polydimethylsiloxane) fiber (20 mm × o.d. 1.1 mm, *df* = 100 μm; Shimadzu).
4. NIST library (NIST/EPA/NIH EI-MS Library).

3 Methods

3.1 Samples Preparation

1. Place three beans (raw/fermented/roasted/nibs) in a polycarbonate tube with a stainless-steel ball and cool in a container with liquid nitrogen for 5 min. In case of samples in the form of cocoa liquor or mass, cocoa cake, and chocolate bar, around 50–100 mg can be used as samples (*see* **Note 1**).
2. In the case of cocoa liquor, if the sample is stored in the fridge, it will be in solid form.
3. Take the cocoa liquor out of the fridge and thaw it at room temperature.
4. Clean the surface of condensed water or dew before weighing or proceeding with the next steps.
5. Place some cocoa liquor solid to fill about half of a polycarbonate tube, then add a stainless-steel ball on top of the sample.
6. Crushed the samples into powder using Multi-Beads Shocker at 2000 rpm for 60 s.
7. Add more cycles of cooling (**step 1**) and crushing (**step 2**) if samples are not homogenized or become paste (*see* **Note 2**).
8. Remove the stainless-steel ball from the polycarbonate tube (*see* **Note 3**).
9. Store the homogenized samples in a −30 °C freezer till analysis.

3.2 Nonvolatile Analysis

3.2.1 Sample Extraction

1. Prepare the extraction solvent consisting of methanol, chloroform, and ultrapure water in a ratio of 2.5:1:1 (v/v/v) and an internal standard (ribitol 0.1 mg/mL) (*see* **Note 4**).
2. Place five milligrams ($n = 3$) of the homogenized sample powder in a 2 mL microcentrifuge tube (*see* **Notes 5** and **6**).
3. Add 1000 μL of extraction solvent into a 2 mL microcentrifuge tube. Prepare microcentrifuge tubes for blank samples as well.

These blank samples of microcentrifuge tubes contain only extraction solvent (no powdered samples).

4. Vortex and incubate all samples in a thermomixer (1200 rpm, 37 °C for 30 min).
5. Centrifuge for 3 min at 4 °C, 10,000 rpm.
6. Collect 600 μL of the supernatants and transfer to a new 1.5 mL microcentrifuge tube.
7. Add 300 μL of ultrapure water and vortex.
8. Centrifuge samples again for 3 min at 4 °C, 10,000 rpm.
9. Separate supernatants into two parts. Transfer 200 μL of supernatant to a new 1.5 mL microcentrifuge tube and collect 200 μL of supernatant in each sample for quality control (QC) samples in a 15 mL centrifuge tube, then vortex.
10. Transfer 200 μL of mixed supernatant from all samples into a new 1.5 mL microcentrifuge tube for making QC samples.
11. Evaporate all the samples to dryness with a vacuum concentrator for 2 h at room temperature (±25 °C)—until no water is left (*see* **Note 7**).
12. Store the samples at −30 °C freezer till analysis.

3.2.2 Derivatization

1. Thaw frozen samples at room temperature (±25 °C) (*see* **Note 5**).
2. Freshly prepare methoxyamine hydrochloride (20 mg/mL) in pyridine and mix vigorously until dissolved.
3. Add 100 μL of methoxyamine hydrochloride solvent to the dried samples, vortex, and flash down the liquid.
4. Incubate the samples in a thermomixer at 30 °C for 90 min and 1200 rpm.
5. Add 50 μL of MSTFA to each sample.
6. Incubate the resulting solution at 37 °C for 30 min and 1200 rpm.
7. Transfer 100 μL of reaction solution into a GC-MS vial and cap firmly. Samples are ready to inject.

3.2.3 GC-MS Parameter

1. The injection amount is 1 μL with a split ratio of 1:25.
2. The injection temperature is set to 270 °C.
3. Set the helium carrier gas flow rate to 1.12 mL/min.
4. The column temperature gradient was initially set at 80 °C for 2 min, then increased by 15 °C/min until reaching the final temperature of 330 °C—hold for 6 min.
5. The transfer line temperature is set to be maintained at 310 °C, while the ion source temperature is set at 280 °C.

6. The electron ionization source operates at −70 eV.
7. The mass spectrometer operates in scan mode at a rate of 10,000 μ/s within a mass range of 85–500 *m/z*.
8. A standard alkane mixture (C8–C40) is set to be injected first before the samples.

3.3 Volatile Analysis

3.3.1 Sample Extraction

1. Approximately 1.5 g of cocoa liquor was added to a 20 mL SPME vial.
2. QC samples are prepared by adding 1.5 g of each sample to a baker's glass, homogenized using a spatula, and then placed 1.5 g into a 20 mL vial.
3. Sample extraction was performed at 60 °C for 30 min with agitation using an autosampler (AOC-6000 by Shimadzu) fitted with a DVB/CAR/PDMS fiber.
4. Following extraction, the fiber was placed into the GC inlet, and the analyte was injected via thermal desorption at 250 °C for 2 min.

3.3.2 GC-MS Parameter

1. The analyte desorbed from the SPME fibers was injected at a split ratio of 25:1 (v/v).
2. A helium carrier gas with a linear velocity of 40.7 cm/s.
3. The column oven temperature was initially set at 40 °C for 5 min, then gradually increased to 250 °C at a rate of 3 °C/min, and maintained at 250 °C for 45 min.
4. The MS was operated in scan mode (m/z 24–350).
5. The ion source and interface temperatures were set at 250 °C.
6. Fatty acid ethyl esters (C4–C24) was injected at the beginning of the analysis.

4 Notes

1. During the cooling process, liquid nitrogen will start to boil after polycarbonate tubes containing stainless-steel balls and samples are placed inside due to temperature differences. Close the container and wait for 5 min. If liquid nitrogen still boils, please wait longer until it boils off to make sure that the samples are already cooled. Take out only when the polycarbonate tubes are ready to be put inside the Multi-Beads Shocker and the crushing process starts.
2. Depending on the fat content, one cycle of cooling and crushing might not be enough. If samples are assumed to have high-fat content, reduce the crushing time to below 60 s, then cool again in liquid nitrogen before crushing again.

3. Stainless steel balls can be removed with the help of a magnet.
4. It is recommended that the extraction solvent be prepared freshly to avoid oxidation of chloroform.
5. Samples taken from a −30 °C freezer should be thawed and the surface cleaned from condensed water or dew prior to weighing or next activities.
6. Weighing must be done in an air-conditioned room and handled while using gloves. Powdered samples of raw and fermented cocoa beans with high-fat content can become paste if touched by a human hand.
7. Ensure that the temperature stays at ±25 °C and there is no water left inside the tube. Add more time if the tube still contains extraction solvent. This step is important to prevent matrix effect.

References

1. Akoa SP, Boulanger R, Effa Onomo P, Lebrun M, Ondobo ML, Lahon MC, Ntyam Mendo SA, Niemenak N, Djocgoue PF (2023) Sugar profile and volatile aroma composition in fermented dried beans and roasted nibs from six controlled pollinated Cameroonian fine-flavor cocoa (*Theobroma cacao* L.) hybrids. Food Biosci 53. https://doi.org/10.1016/j.fbio.2023.102603
2. ICCO (2017) What is fine flavour cocoa?
3. Bacca-Villota P, Acuña-García L, Sierra-Guevara L, Cano H, Hidalgo W (2023) Untargeted metabolomics analysis for studying differences in high-quality Colombian cocoa beans. Molecules 28(11). https://doi.org/10.3390/molecules28114467
4. Hanifah A, Firmanto H, Putri SP, Fukusaki E (2022) Unique metabolite profiles of Indonesian cocoa beans from different origins and their correlation with temperature. J Biosci Bioeng 134(2):125–132. https://doi.org/10.1016/j.jbiosc.2022.05.001
5. Escobar S, Santander M, Zuluaga M, Chacón I, Rodríguez J, Vaillant F (2021) Fine cocoa beans production: tracking aroma precursors through a comprehensive analysis of flavor attributes formation. Food Chem 365. https://doi.org/10.1016/j.foodchem.2021.130627
6. Herrera-Rocha F, León-Inga AM, Aguirre Mejía JL, Rodríguez-López CM, Chica MJ, Wessjohann LA, González Barrios AF, Cala MP, Fernández-Niño M (2024) Bioactive and flavor compounds in cocoa liquor and their traceability over the major steps of cocoa post-harvesting processes. Food Chem 435. https://doi.org/10.1016/j.foodchem.2023.137529
7. Colonges K, Jimenez JC, Saltos A, Seguine E, Loor Solorzano RG, Fouet O, Argout X, Assemat S, Davrieux F, Cros E, Lanaud C, Boulanger R (2022) Integration of GWAS, metabolomics, and sensorial analyses to reveal novel metabolic pathways involved in cocoa fruity aroma GWAS of fruity aroma in Theobroma cacao. Plant Physiol Biochem 171:213–225. https://doi.org/10.1016/j.plaphy.2021.11.006
8. Balcázar-Zumaeta CR, Castro-Alayo EM, Cayo-Colca IS, Idrogo-Vásquez G, Muñoz-Astecker LD (2023) Metabolomics during the spontaneous fermentation in cocoa (*Theobroma cacao* L.): an exploraty review. Food Res Int 163. https://doi.org/10.1016/j.foodres.2022.112190. Elsevier Ltd
9. de Sousa Dias AL, Meudec E, Verbaere A, Lair S, Boulet JC, Sommerer N (2024) Cocoa bean metabolomics reveals polyphenols as potential markers relating to fine dark chocolate color shades. Front Nutr 11. https://doi.org/10.3389/fnut.2024.1467282
10. Afifah EN, Sari IA, Susilo AW, Malik A, Fukusaki E, Putri SP (2024) Characterization of fine-flavor cocoa in parent-hybrid combinations using metabolomics approach. Food Chem X 24. https://doi.org/10.1016/j.fochx.2024.101832
11. Fang Y, Li R, Chu Z, Zhu K, Gu F, Zhang Y (2020) Chemical and flavor profile changes of cocoa beans (*Theobroma cacao* L.) during primary fermentation. Food Sci Nutr 8(8):

4121–4133. https://doi.org/10.1002/fsn3.1701
12. Qin XW, Lai JX, Tan LH, Hao CY, Li FP, He SZ, Song YH (2017) Characterization of volatile compounds in Criollo, Forastero, and Trinitario cocoa seeds (*Theobroma cacao* L.) in China. Int J Food Prop 20(10):2261–2275. https://doi.org/10.1080/10942912.2016.1236270
13. Velásquez-Reyes D, Rodríguez-Campos J, Avendaño-Arrazate C, Gschaedler A, Alcázar-Valle M, Lugo-Cervantes E (2023) Forastero and Criollo cocoa beans, differences on the profile of volatile and non-volatile compounds in the process from fermentation to liquor. Heliyon 9(4). https://doi.org/10.1016/j.heliyon.2023.e15129
14. Moreno-Rojas JM, Yadira Erazo Solorzano C, Tuárez García DA, Pereira-Caro G, Ordóñez Díaz JL, Muñoz-Redondo JM, Rodríguez-Solana R (2023) Impact of the pre-drying process on the volatile profile of on-farm processed Ecuadorian bulk and fine-flavour cocoa varieties. Food Res Int 169. https://doi.org/10.1016/j.foodres.2023.112938
15. Kitani Y, Putri SP, Fukusaki E (2022) Investigation of the effect of processing on the component changes of single-origin chocolate during the bean-to-bar process. J Biosci Bioeng 134(2):138–143. https://doi.org/10.1016/j.jbiosc.2022.05.007
16. Bastos VS, Uekane TM, Bello NA, de Rezende CM, Flosi Paschoalin VM, Del Aguila EM (2019) Dynamics of volatile compounds in TSH 565 cocoa clone fermentation and their role on chocolate flavor in Southeast Brazil. J Food Sci Technol 56(6):2874–2887. https://doi.org/10.1007/s13197-019-03736-3
17. Amalia F, Aditiawati P, Yusianto, Putri SP, Fukusaki E (2021) Gas chromatography/mass spectrometry-based metabolite profiling of coffee beans obtained from different altitudes and origins with various postharvest processing. Metabolomics 17(7). https://doi.org/10.1007/s11306-021-01817-z
18. Jumhawan U, Putri SP, Yusianto, Bamba T, Fukusaki E (2016) Quantification of coffee blends for authentication of Asian palm civet coffee (Kopi Luwak) via metabolomics: a proof of concept. J Biosci Bioeng 122(1):79–84. https://doi.org/10.1016/j.jbiosc.2015.12.008
19. Putri SP, Ikram MMM, Sato A, Dahlan HA, Rahmawati D, Ohto Y, Fukusaki E (2022) Application of gas chromatography-mass spectrometry-based metabolomics in food science and technology. J Biosci Bioeng 133(5):425–435. https://doi.org/10.1016/j.jbiosc.2022.01.011. Elsevier B.V
20. Rojas SM, Chejne F, Ciro H, Montoya J (2020) Roasting impact on the chemical and physical structure of Criollo cocoa variety (Theobroma cacao L). J Food Process Eng 43(6). https://doi.org/10.1111/jfpe.13400

Chapter 4

GC-MS-Based Metabolite Profiling Method for Young Coconut

Mary Faith Yamballa Adan

Abstract

The chemical profile of young coconut (*Cocos nucifera* L.) is important in defining its physical, nutritional, and sensory qualities. Gas chromatography-mass spectrometry (GC-MS) is emerging as a powerful tool for investigating the chemical components of young coconuts. Since young coconuts contain high lipids and diverse metabolites, reliable analysis depends on thoughtful and consistent sample preparation. This chapter offers a comprehensive protocol for GC-MS-based metabolomic analysis of both liquid and solid endosperm of young coconuts, detailing key procedures such as sample preparation, extraction, and derivatization. It explains the rationale behind the method and provides practical guidance for performing GC-MS analytical workflows. Solving the challenges due to metabolite diversity and matrix composition, the protocol enables reliable compound identification even across various research studies. These methods can help researchers get consistent and comparable results to enhance the resolution of metabolite profiling and compound identification in young coconut samples to facilitate a deeper understanding of how their chemical components affect their physical, nutritional, and sensorial qualities.

Key words *Cocos*, Young Coconut, Gas chromatography, Mass spectrometry, Sample preparation

1 Introduction

Young coconut, derived from the coconut palm (*Cocos nucifera* L.), is increasingly sought after for its unique flavor and nutritional benefits. Its popularity has seen a resurgence as global consumers embrace it as a health-promoting food, fueling economic growth across Southeast Asia [1]. The global market value for coconut products is reported at approximately $13 billion in 2019 and projected to increase to $31.1 billion in 2026 [2]. Predominantly cultivated in regions such as Southeast Asia and Melanesia, the export volumes are continuously increasing to meet the rising demand for healthy and natural young coconuts among consumers in developed regions [3, 4].

The quality attributes of young coconuts have become pivotal for the exportation market, prompting research efforts to

Sastia Prama Putri (ed.), *Gas Chromatography-Mass Spectrometry: Based Metabolomics in Food Science*, Methods and Protocols in Food Science, https://doi.org/10.1007/978-1-0716-5108-7_4,

characterize and assess their physical, compositional, nutritional, and sensorial qualities. Research investigations span the impact of varietal differences [5, 6], storage conditions [7], developmental stages [8–10], and preservation processes [11] on these quality attributes of young coconut. Notably, its chemical components, including volatiles, sugars, and amino acids, play a critical role in defining its quality attributes, making young coconut an increasingly prominent focus of metabolomic-based studies [12, 13].

In plant metabolomic studies, the complex network of fruit metabolites can be investigated using several highly sensitive and selective analytical techniques. Among these techniques, gas chromatography-mass spectrometry (GC-MS) is commonly used to investigate fruit metabolites due to its advantageous characteristics, such that it can measure a wide range of small molecules with good reproducibility and robustness. It also has a readily available library and databases used in metabolite identification [14–16].

However, processing coconut samples for GC-MS has challenges because they contain high lipid content along with various metabolites that vary in polarity and structure [13]. To analyze young coconut samples, an appropriate extraction and derivatization method must be employed so that the compounds stay stable and can be measured accurately. Although many studies have been published using GC-MS to analyze coconut samples, detailed descriptions of the methodologies are not reported. To ensure reproducibility, protocols with detailed procedural workflows encompassing young coconut sample preparation, derivatization, and analysis conditions are essential for robust GC-MS-based metabolomics [17]. Hence, this chapter provides a clear and practical protocol for preparing, extracting, and derivatizing metabolites from young coconut flesh and water for GC-MS analysis. It outlines each step in detail while explaining the reasoning behind key procedures and giving useful tips for improving speed, consistency, and accuracy. The method is designed to tackle the unique chemical complexity of young coconuts, having diverse metabolite structures. Researchers in food science, nutrition, and metabolomics can apply this protocol to get comparable and reproducible results through different kinds of studies.

2 Materials

2.1 Sample and Reagent

1. Young coconut fruit.
2. Liquid nitrogen.
3. Pure water (resistivity between 0.1 and 15 megaohms per centimeter, MΩ-cm).
4. Hydrogen gas.

5. Methanol (LC-MS grade).
6. Chloroform (LC-MS grade).
7. Ultrapure water (resistivity >15 megaohms per centimeter, MΩ-cm).
8. Ribitol, as internal standard (purity ≥99.0%).
9. Methoxyamine hydrochloride (purity, 98%).
10. Ultrapure pyridine (GC Capillary grade, 99.5+%).
11. *N*-methyl-*N*-(trimethylsilyl)-trifluoroacetamide (MSTFA).
12. Retention time index standard mixture: C9–C40 alkane standard mixture (standard grade for GC).

2.2 Laboratory Supplies

1. Spatula.
2. 1-L beaker (to collect young coconut water).
3. 50-mL conical centrifuge tubes (for storing collected samples).
4. Calibrated 1000 μL and 200 μL micropipette.
5. Freeze dryer (example: TAITEC freeze dryer VD-800R).
6. Multi-beads shocker (example: Yasui Kikai MB3000 series).
7. 50-mL polycarbonate tubes (for multi-bead shocker).
8. Vacuum sealer and appropriate vacuum-sealer bag.
9. Freezer-safe marker pen for labeling the samples.
10. Freezer gloves for handling frozen objects.
11. −80 °C freezer to store samples.
12. 2.0-mL microcentrifuge tubes.
13. 1.5-mL microcentrifuge tubes.
14. 15-mL conical centrifuge tubes (for pooling QC samples).
15. 4 °C refrigerator to store layered supernatant (optional).
16. Analytical balance (capable of measuring 5 mg).
17. Vortex mixer.
18. Refrigerated centrifuge.
19. Thermomixer.
20. Speed vacuum concentrator.
21. GC-MS vials equipped with 200 μL insert.

2.3 GC-MS Equipment

1. GC-MS system: GCMS-QP2010 Ultra (Shimadzu, Kyoto, Japan) equipped with an AOC-20i/s autoinjector (Shimadzu).
2. GC capillary column: InertCap 5 MS/NP column (GL Sciences).
3. NIST mass spectral library (NIST/EPA/NIH EI-MS Library).
4. If available, in-house GC-MS-5MP Library (RIKEN).

3 Methods

3.1 Sample Preparation for Young Coconut Water

1. Collect the young coconut fruit and wash thoroughly with clean water. Dry the fruit using a clean cloth. Remove and peel the outer layer (exocarp) and most of the husk (mesocarp), being careful not to puncture the endocarp using a clean knife. Use a clean young coconut opening drill tool and drain all the coconut water into a clean beaker (*see* **Note 1**).
2. Distribute the collected coconut water into new 50-mL conical centrifuge tubes pre-labeled with the appropriate sample name or code and date of collection.
3. Depending on the objective of the analysis, it can immediately be used for analysis (*see* **Note 2**), or it can be quenched using liquid nitrogen (*see* **Note 3**) and stored in a −80 °C freezer before extraction and further analysis.
4. One milliliter (1 mL) of young coconut water will be used per replicate for the analysis. If samples were stored, thaw them properly before using them further (*see* **Note 5**).

3.2 Sample Preparation for Young Coconut Flesh

1. Collect the young coconut fruit and wash thoroughly with clean water. Dry the fruit using a clean cloth. Remove and peel the outer layer (exocarp) and most of the husk (mesocarp), being careful not to puncture the endocarp using a clean knife. Cut the fruit into two parts using a clean knife (*see* **Note 1**) and drain all the coconut water into a clean container to discard or to keep if needed for further analysis (refer to Subheading 3.1).
2. Scrape off all the flesh using a clean spatula, then distribute the flesh to 50-mL polycarbonate tubes for the multi-beads shocker (pre-labeled with the appropriate sample name or code and date of collection), filling only two-thirds of the tube, and close their caps tightly.
3. Immediately quench the collected flesh samples by submerging the whole tube in liquid nitrogen and wait until the bubbling disappears (*see* **Notes 3** and **4**).
4. Change the caps of the polycarbonate tubes to parafilm and make several small holes in the parafilm. Freeze-dry the coconut flesh for 48 h using a TAITEC freeze dryer VD-800R.
5. After freeze-drying, remove the parafilm and put a bead inside each tube before tightly closing them with their original caps. Then, submerge the whole tube with samples and beads in liquid nitrogen and wait until the bubbling disappears (*see* **Notes 3** and **4**).
6. Crush the freeze-dried flesh sample using a multi-beads shocker for 20 s at 2000 rpm. Pool and mix all the samples in

a single vacuum-sealer bag and vacuum-seal the bag for storage.

7. Immediately use for analysis or store at −80 °C until analysis.
8. Five milligrams of the coconut flesh powder will be used per replicate. If samples were stored, thaw them properly before using them further (*see* **Note 6**).

3.3 Sample Extraction

1. Prepare the appropriate amount of extraction solvent consisting of methanol, water, and chloroform (ratio of 5:2:2 v/v/v) with ribitol (100 μg/mL) as the internal standard (*see* **Notes 7** and **8**). Work in a fume hood (*see* **Note 9**).
2. Weigh 5 mg of coconut flesh powder in a 2-mL microtube for each sample replicate or place 1 mL of the prepared coconut water in a 2-mL microtube for each sample replicate. Prepare 2-mL microtubes for blank samples as well. Label all microtubes properly.
3. Add 1 mL of the extraction solvent (methanol, water, and chloroform in a ratio of 5:2:2 v/v/v) to each sample replicate in 2-mL microtubes and vortex for 10 s.
4. Incubate the mixture in a thermomixer set at 1200 rpm and 37 °C for 30 min (*see* **Note 10**).
5. Centrifuge the samples at 4 °C and 10,000 rpm for 3 min to separate the suspended solids from the liquid (*see* **Note 10**).
6. Prepare 300 μL of ultrapure water in 1.5 mL tubes and add 600 μL of the supernatant from the previous step (**step 5**).
7. Centrifuge the samples at 4 °C and 10,000 rpm for 3 min (*see* **Note 10**).
8. Transfer 50 μL of the aqueous layer (*see* **Notes 11** and **12**) into new, pre-labeled 1.5-mL microtubes for the samples and blanks. Collect an additional 50 μL from the aqueous layer of each sample and transfer into a new, pre-labeled 15-mL conical centrifuge tube for the pooled quality control (QC) samples.
9. Redistribute the pooled quality control samples and place 50 uL each in separate 1.5-mL microtubes for each QC sample needed for the analysis (*see* **Note 13**).
10. Open the caps of each 1.5-mL microtube containing the collected aqueous layer from the samples and blanks (from **step 8**) and QC samples (from **step 9**). Then, close the tubes using premade perforated caps (*see* **Note 14**) without cutting the original caps from the tubes.
11. Centrifuge the tubes closed with perforated caps at 25 °C and 1500 rpm for 60 min using a centrifugal concentrator (TAITEC, Saitama, Japan) under vacuum to obtain the extract.

12. After centrifugal concentration, collect all the microtubes and remove the perforated caps before properly closing them using their original caps.
13. Store the tubes at −80 °C until further use.

3.4 Sample Derivatization

1. Take out the dried extracts from the freezer and thaw all microtubes at room temperature for 15 min or as needed (*see* **Note 15**).
2. Prepare a fresh methoxyamine solution by dissolving methoxyamine hydrochloride at 20 mg/mL in pure pyridine. Work in a fume hood (*see* **Note 9**).
3. Once the dried extracts have reached room temperature, add 100 μL of the prepared methoxyamine hydrochloride (20 mg/mL in pyridine) from **step 2** to each of the samples, QC, and blank tubes, then immediately close the tubes.
4. Incubate the mixture in a thermomixer at 30 °C and 1200 rpm for 90 min (*see* **Note 10**).
5. Remove tubes from the thermomixer and spin down for 10 s to collect the drops on the walls and caps of the tubes.
6. Add 50 μL of *N*-methyl-*N*-trimethylsilyl-trifluoroacetamide (MSTFA) to each sample, QC, and blank microtubes.
7. Incubate the mixture in a thermomixer at 37 °C and 1200 rpm for 30 min (*see* **Note 10**).
8. Transfer the derivatized samples to GC vials for GC-MS analysis.

3.5 Data Acquisition by GC-MS

1. Conduct the analysis using a GCMS-QP2010 Ultra (Shimadzu, Kyoto, Japan) equipped with an AOC-20i/s-autoinjector (Shimadzu). Program the injections to include syringe washing steps before and after each injection. Autotune the instrument according to the manufacturer's instructions.
2. Perform chromatography using an InertCap 5MS/NP column (length 35 m, internal diameter (I.D.) 0.18 mm, and diameter film/film thickness (*df*) = 0.18 μm) (GL Sciences, Tokyo) (*see* **Note 16**).
3. Inject 1 μL of each sample into the GC at a temperature of 230 °C using a split ratio of 25:1.
4. Set the linear velocity of the carrier gas (H_2, hydrogen gas) at 39.0 cm/s.
5. Maintain the column temperature at 80 °C for 4 min, then increase to 330 °C at a rate of 15 °C/min, and finally hold it at 330 °C for 8 min.
6. Set the interface and ion source temperatures at 310 and 280 °C, respectively.

7. Set the filament bias voltage at 70.0 V for producing ions via electron ionization (EI).
8. Register the spectra over a mass range m/z 85–500 for 0.15 s.
9. Inject a standard alkane mixture of C_9–C_{40} into the GC at the beginning of the analysis to calculate the retention index (RI) that will be used to identify peaks.

4 Notes

1. Young coconut opener drills or tools are available online or in specialized shops. To prevent contamination of the samples, make sure to clean the beakers and opening tools thoroughly by cleaning with 95% ethanol and rinsing with pure water thrice.
2. It is better to subject the samples to analysis immediately to prevent any changes in the metabolite composition, but depending on the objective of the analysis, it can be quenched using liquid nitrogen and stored at −80 °C until further analysis.
3. Immediately quench the collected young coconut samples by submerging the tightly closed 50-mL conical centrifuge tubes (or polycarbonate tubes) containing the samples in liquid nitrogen, and wait until the bubbling disappears to make sure that the samples are properly frozen.
4. Take out the polycarbonate tubes containing samples (and beads) from the liquid nitrogen only when they are going to be freeze-dried using a freeze dryer or homogenized using a multi-bead shocker. Otherwise, store them at a −80 °C freezer and take them out immediately before they are going to be processed.
5. To thaw the water samples, take them out of the −80 °C freezer and place them in an ice bath until the water is fully thawed.
6. To thaw the powdered flesh samples, take them out from the −80 °C freezer and place them at room temperature for 15–30 min until the tubes reach room temperature. Longer thawing time has not been investigated.
7. The volume of the mixed solvent (composed of methanol, chloroform, and ultrapure water at a ratio of 5:2:2) can be calculated considering that each tube of samples and blanks will use 1000 μL and, optionally, we add at least 4000 μL of extra volume. The calculation will be as follows:

   Mixed solvent total volume = [1000 μL × (number of sample tubes + number of blank tubes)] + 4000 μL as extra volume.

8. The internal standard, ribitol, will be included in the mixed solvent. Prepare a 20 mg/mL ribitol stock concentration and use the necessary amount for the mixed solvent. The final concentration of ribitol in the mixed solvent should be 100 μg/mL. Order of mixing: Methanol + water → ribitol → chloroform.
9. Chemical reagents are hazardous and should be handled in a fume hood with gloves and proper safety goggles.
10. Remember to pre-set the thermomixer or refrigerated centrifuge to the necessary temperature, rotation speed, and incubation time at least 10 min before use (or before adding the appropriate reagent to the microtubes).
11. After centrifugation, there will be two separate layers in the tube. The top layer is the aqueous layer, which contains relatively hydrophilic compounds, and the bottom layer is the chloroform layer, which contains hydrophobic compounds.
12. The supernatant after centrifugation, containing the aqueous and chloroform layers, can be stored at 4 °C for 1 day, in case more extract is needed later. Storage of more than 1 day has not been investigated.
13. To determine how many QC samples are needed for the analysis, use the following formula: Number of quality control (QC) samples = number of samples 5 + 4.
14. Perforated caps are made by cutting clean 1.5-mL microtube caps and making 3–4 holes on it using a strong sharp object (such as a big pin). After using, the perforated caps can be washed by submerging in methanol and ultrasonicating for 10 min, then rinsing with ultrapure water twice before drying in an oven at 60 °C overnight.
15. Thawing time of more than 30 min has not been investigated.
16. Prior to analysis, an inner seal connector and inert silica capillary tube should be prepared at 250 °C for 1 h and then combine with the column setup.

References

1. DebMandal M, Mandal S (2011) Coconut (*Cocos nucifera* L.: Arecaceae): in health promotion and disease prevention. Asian Pac J Trop Med 4(3):241–247. https://doi.org/10.1016/S1995-7645(11)60078-3
2. Zainol FA, Arumugam N, Daud WNW et al (2023) Coconut value chain analysis: a systematic review. Agriculture (Basel) 13(7):1379. https://doi.org/10.3390/agriculture13071379
3. Sáenz-Carbonell L, Montero-Cortés M, Pérez-Nuñez T et al (2016) Somatic embryogenesis in *Cocos nucifera* L. In: Somatic embryogenesis: fundamental aspects and applications. Springer International Publishing, New York, pp 297–318. https://doi.org/10.1007/978-3-319-33705-0_18
4. Siriphanich J, Saradhuldhat P, Romphophak T et al (2011) Coconut (*Cocos nucifera* L.). In: Postharvest biology and technology of tropical and subtropical fruits. Woodhead Publishing,

Cambridge, pp 8–35e. https://doi.org/10.1533/9780857092885.8

5. Zhang W, Chen Y, Yun Y et al (2023) Discrimination and characterization of different coconut water (CW) by their phenolic composition and volatile organic compounds (VOCs) using LC-MS/MS, HS-SPME-GC-MS, and HS-GC-IMS. J Food Sci 88(9):3758–3772. https://doi.org/10.1111/1750-3841.16711
6. Yunindanova MB, Putri SP, Novarianto H et al (2024) Characteristics of kopyor coconut (*Cocos nucifera* L.) using sensory analysis and metabolomics-based approach. J Biosci Bioeng 138(1):44–53. https://doi.org/10.1016/j.jbiosc.2024.02.008
7. Rahmawati D, Adan MFY, Ikram MMM et al (2024) Effect of sodium metabisulfite treatment and storage condition on metabolic profile of young coconut (*Cocos nucifera* L.). J Biosci Bioeng 138(6):515–521. https://doi.org/10.1016/j.jbiosc.2024.08.002
8. Kumar M, Saini SS, Agrawal PK et al (2021) Nutritional and metabolomics characterization of the coconut water at different nut developmental stages. J Food Compos Anal 96. https://doi.org/10.1016/j.jfca.2020.103738
9. Kumar M, Agrawal PK, Roy P et al (2022) GC-MS-based metabolomics reveals dynamic changes in the nutritionally important metabolites in coconut meat during nut maturation. J Food Compos Anal 114. https://doi.org/10.1016/j.jfca.2022.104869
10. Jirapong C, Wongs-Aree C, Noichinda S et al (2015) Assessment of volatile and non-volatile organic compounds in the liquid endosperm of young 'Nam Hom' coconut (Cocos nucifera L.) at two stages of maturity. J Hortic Sci Biotechnol 90(5):477–482. https://doi.org/10.1080/14620316.2015.11668703
11. Cunha AG, Alves Filho EG, Silva LMA et al (2020) Chemical composition of thermally processed coconut water evaluated by GC–MS, UPLC-HRMS, and NMR. Food Chem 324:126874. https://doi.org/10.1016/j.foodchem.2020.126874
12. Zhang Y, Kan J, Liu X et al (2024) Chemical components, nutritional value, volatile organic compounds and biological activities in vitro of coconut (*Cocos nucifera* L.) water with different maturities. Foods 13(6):863. https://doi.org/10.3390/foods13060863
13. Guo H, Lai J, Li C et al (2022) Comparative metabolomics reveals key determinants in the flavor and nutritional value of coconut by HS-SPME/GC-MS and UHPLC-MS/MS. Metabolites 12(8):691. https://doi.org/10.3390/metabo12080691
14. Putri SP, Ikram MMM, Sato A et al (2022) Application of gas chromatography-mass spectrometry-based metabolomics in food science and technology. J Biosci Bioeng 133(5):425–435. https://doi.org/10.1016/j.jbiosc.2022.01.011
15. Pott DM, Vallarino JG, Osorio S (2020) Metabolite changes during postharvest storage: effects on fruit quality traits. Metabolites 10(5):187. https://doi.org/10.3390/metabo10050187
16. Obata T, Fernie AR (2012) The use of metabolomics to dissect plant responses to abiotic stresses. Cell Mol Life Sci 69(19):3225–3243. https://doi.org/10.1007/s00018-012-1091-5
17. Shim M, Jeong Y, Lee DK (2025) Identification and profiling of primary metabolites through GC-MS and associated data processing. In: Natural product isolation and identification: methods and protocols. Springer US, New York, pp 99–109. https://doi.org/10.1007/978-1-0716-4350-1_8

Chapter 5

GC-MS-Based Metabolite Profiling Method for Coffee

Fitri Amalia and Yulianti

Abstract

Coffee contains numerous both nonvolatile and volatile compounds, which contribute to its widely known flavor and aroma. Therefore, analyzing the composition of these compounds is essential for coffee quality research and enhancing the understanding of their roles in coffee quality. The metabolomics approach provides metabolite profiling, covering a wide range of compounds, and has become a common method in coffee research. The chapter details the gas chromatography-mass spectrometry (GC-MS)-based metabolomics workflow used to analyze both nonvolatile and volatile compounds in coffee. Two separate subchapters are provided to focus on the specific methodologies for sample preparation, extraction, and derivatization needed for each compound type.

Key words Coffee, Derivatization, Extraction, GC/MS, Nonvolatile analysis, Volatile analysis

1 Introduction

Coffee consumption has increased steadily over the past five decades, making it one of the most important commodities traded globally [1]. The pleasant aroma and flavor of coffee are the main reasons behind its popularity and are essential attributes for determining coffee quality [2, 3]. These sensory characteristics are closely linked to the biochemical composition of coffee beans. A wide range of nonvolatile and volatile compounds contribute to coffee flavor and aroma, formed through a complex series of chemical transformations from seed to cup [4]. Green coffee beans contain nonvolatile compounds such as polysaccharides, free amino acids, proteins, and lipids, which serve as precursors to numerous volatile compounds formed during roasting [4–6]. Approximately 200 volatile compounds have also been identified in green coffee beans, including hydrocarbons, aldehydes, alcohols, ketones, esters, acids, lactones, phenols, sulfur compounds, and furans, which contribute green and musty notes [4–6].

Sastia Prama Putri (ed.), *Gas Chromatography-Mass Spectrometry: Based Metabolomics in Food Science*, Methods and Protocols in Food Science, https://doi.org/10.1007/978-1-0716-5108-7_5,

After roasting, the number of volatile compounds increases to over 1000, including ketones, aldehydes, alcohols, pyrazines, esters, furans, nitrogen-containing compounds, acids, and volatile phenolics [7].

The composition of nonvolatile and volatile compounds in both green and roasted coffee beans is influenced by several factors. In green beans, the chemical profile is affected by geographical origin and postharvest processing methods such as fermentation, drying, and storage [4, 8–12]. In roasted beans, the chemical composition is primarily influenced by roasting conditions and brewing methods, which significantly impact coffee aroma quality [13–15]. Therefore, to better understand these influences on coffee quality, the investigation of nonvolatile and volatile compounds in coffee has become a major focus in coffee research.

Variations in coffee chemical components can be effectively analyzed using metabolomics approaches, which have shown promising results in recent studies [8, 10, 11, 16–19]. Gas chromatography coupled with mass spectrometry (GC-MS) is a widely used method for analyzing both volatile and nonvolatile compounds in coffee due to its high sensitivity and ability to detect compounds at low concentrations [20]. For nonvolatile compound analysis, derivatization is often required to enhance compound volatility, typically through oximation and silylation processes [21, 22]. For volatile compound analysis, solid-phase microextraction (SPME)—especially headspace SPME (HS-SPME)—is commonly used. This technique utilizes a fiber coated with a polymer, such as divinylbenzene/carboxen/polydimethylsiloxane (DVB/CAR/PDMS), to rapidly adsorb and concentrate analytes, which are then thermally desorbed for GC-MS analysis [23].

In this chapter, detailed GC-MS-based analytical methods are presented to comprehensively evaluate nonvolatile and volatile compounds in coffee beans. Separate sections describe specific techniques for sample preparation, extraction, and derivatization. GC-MS enables the reliable identification and annotation of compounds, providing insights into chemical changes during processing stages, such as fermentation, drying, roasting, brewing, and storage. This technology facilitates the detection of both major and minor compounds, which may significantly affect coffee quality. Ultimately, the insights from GC-MS analysis can support the optimization of coffee processing methods and enhance the overall quality of the final product.

2 Materials

2.1 Common Materials

1. Green coffee beans.
2. Roasted coffee beans.
3. Brewed coffee.

2.2 Nonvolatile Analysis

2.2.1 Chemicals

1. Methanol (99.8%).
2. Ultrapure water.
3. Chloroform (99%).
4. Ribitol standard.
5. Methoxyamine hydrochloride.
6. Pyridine (99.5%).
7. N-Methyl-N-(trimethylsilyl) trifluoroacetamide (MSTFA).
8. Alkane mixture (C9–C40) standard.
9. Liquid nitrogen.

2.2.2 Laboratory Supplies

1. Grinding tube 50 mL.
2. Metal cone.
3. Eppendorf tube 2 mL.
4. Eppendorf tube 1.5 mL.
5. Eppendorf tube 15 mL.
6. GC/MS vials with cap.
7. Amber glass vial/bottle 100 mL.
8. Micropipette tips.
9. InertCap 5MS/NP (0.25 μm × 30 m × 0.25 mm i.d.) fused silica capillary column (GL Science, Inc., Tokyo, Japan).

2.2.3 Laboratory Equipment and Instrument

1. Micropipette.
2. Spatula.
3. CryoTank.
4. Multi-beads shocker.
5. Thermomixer.
6. Vortex.
7. Centrifuge.
8. Vacuum evaporator.
9. Freeze-dryer.
10. GCMS-QP2010 Ultra (Shimadzu, Kyoto, Japan).
11. GCMS-TQ8030 (Shimadzu, Kyoto, Japan).
12. AOC-20i/s autosampler (Shimadzu, Kyoto, Japan).

2.2.4 Software for Data Analysis

1. GC-MS Solution software package (Shimadzu, Kyoto, Japan).
2. Reifycs Abf Converter (https://www.reifycs.com/AbfConverter/).
3. MS-DIAL [24].

4. GL-Science DB Library (InertCap 5MS-NP, Kovats RI, 494 records; http://prime.psc.riken.jp/Metabolomics_Software/MS-DIAL/GCMS%20DB_InertCap%205MS-NP_GLscience.msp).
5. SIMCA-P.
6. R Studio.

2.3 Volatile Analysis

2.3.1 Chemicals

1. Liquid nitrogen.
2. 3-heptanone (internal standard) (*see* Subheading 4.2, **Note 1**).
3. Alkane (C9–C29).

2.3.2 Laboratory Supplies

1. Micropipette tips.
2. SPME vial 22 mL.
3. 2 cm DVB/CAR/PDMS SPME fiber.
4. DB-Wax column (30 m × 250 μm × 0.25 μm).

2.3.3 Laboratory Equipment and Instrument

1. Micropipette.
2. GC (Agilent 7890A, California, USA).
3. MS detector (Agilent 595975 XL EI/CI).

2.3.4 Software for Data Analysis

1. NIST 14 Library (National Institute of Standards and Technology).
2. XLSTAT.
3. SIMCA-P.
4. R Studio.

3 Methods

3.1 Nonvolatile Analysis

3.1.1 Preparing Coffee Bean Powder [8, 10, 17, 24]

1. Put at least 5–10 beans of green coffee beans (raw coffee beans) or roasted coffee beans to into 50 mL grinding tube (Yasui Kikai Co.).
2. Insert the clean metal cone (Yasui Kikai Co.) into the grinding tube.
3. Close the tube tightly.
4. Quench the tube, with coffee beans and metal cone inside, by dipping in liquid nitrogen.
5. Put the quenched tube to a multi-beads shocker (Yasui Kikai Co., Osaka, Japan). Make sure the tube is securely positioned, and close the lid as per the specified machine instructions.
6. Shaken the tube at 2000 rpm for 30 s.

7. Open the lid and check the grinding result inside the tube. It is recommended to repeat the previous **steps 3** and **4** times to produce fine powder (*see* Subheading 4.1, **Note 1**).

3.1.2 Extraction Method [8, 10, 17, 24]

1. Transfer 15 mg of coffee powder into 2 mL Eppendorf tube. It is recommended to prepare three replicates ($n = 3$) for each coffee sample.
2. Prepare an empty tube as a blank, and proceed with all the steps below with samples.
3. Add 1 mL of a mixed solvent (methanol, ultrapure water, and chloroform = 5/2/2 (v/v/v) ratio) containing an internal standard (ribitol 0.2 mg/mL) into the tube.
4. Homogenize and incubate the mixture at 1200 rpm at 37 °C for 30 min.
5. Centrifuge the mixture at 4 °C and 16,000× *g* for 3 min.
6. Transfer 900 μL of supernatant into a new 1.5 mL tube.
7. Add 400 μL of ultrapure water into the tube.
8. Vortex the mixture and centrifuge at 4 °C and 16,000× *g* for 3 min.
9. Transfer 200 μL aqueous phase (the upper layer) to a new 1.5 mL tube and close the tube with a pierced cap.
10. For quality control (QC) purposes, pool 100 μL of the aqueous layer obtained from all samples to a 15 mL Eppendorf tube (blank not included).
11. Homogenize the QC pool mixture, transfer 200 μL extract to a new 1.5 mL Eppendorf tube, and close the tube with a pierced cap.
12. Evaporate the solvent (sample, blank, and QC) using vacuum centrifugation for 50 min.
13. Lyophilize the concentrated mixture overnight.

3.1.3 Derivatization Method [17]

1. Add 100 μL methoxyamine hydrochloride (20 mg/mL in pyridine) to the lyophilized extract.
2. Incubate and homogenize the mixture at 30 °C and 1200 rpm for 90 min.
3. Add 50 μL of MSTFA to the mixture.
4. Incubate and homogenize the mixture again at 37 °C and 1200 rpm for 30 min.
5. The derivatized sample, blank, and QC are ready to be injected into GC-MS (*see* Subheading 4.1, **Note 2**).

3.1.4 Gas Chromatography-Mass Spectrometry Analysis [17]

1. GC/MS analysis is conducted using either GCMS-QP2010 Ultra (Shimadzu, Kyoto, Japan) or GCMS-TQ8030 (Shimadzu, Kyoto, Japan) equipped with an autosampler AOC-20i/s (Shimadzu, Kyoto, Japan).
2. Set the injection mode to split mode (12:1 (v/v)).
3. Set the injection temperature to 270 °C.
4. Use helium as the carrier gas with flow rate of 1.12 mL/min and set the linear velocity at 39 cm/s.
5. Set the column temperature at 80 °C for 2 min, increased by 15 °C/min to 330 °C, and finally kept at 330 °C for 6 min.
6. Set transfer line temperature to 250 °C.
7. Set the ion source temperature to 280 °C.
8. Set the electron ionization at 70 eV.
9. Set the recorded EI scan at 20 scans per second with the mass range *m/z* 85–500.
10. Tune and calibrate the mass spectrometer before sample analysis.
11. Inject the standard alkane mixture (C9–C40) at the beginning of analysis for retention index calculation (*see* Subheading 4.1, **Note 3**).
12. Inject 1 μL the derivatized samples and blank(s) randomly with QC injection for every four to five samples.

3.1.5 Data Preprocessing and Analysis [17]

1. Convert the chromatogram to netCDF format using the GC-MS Solution software package (Shimadzu, Kyoto, Japan).
2. Convert the netCDF file to ABF format using Reifycs Abf Converter.
3. Use ABF format files for peak detection, deconvolution analysis, identification, and alignment using MS-DIAL [25–27].
4. Set the peak detection parameter with a minimum height of 1000 amplitude.
5. Use a linear weight moving average smoothing method.
6. Set the smoothing level to five scans, with an average peak width of 20 scans.
7. Set the σ-window value to 0.5.
8. Set the EI spectra cutoff to 200 amplitudes for the deconvolution analysis.
9. Use the retention time of standard alkane (C9–C40) (*see* Subheading 4.1, **Note 4**) and GL-Science DB Library (InertCap 5MS-NP, Kovats RI,494 records) for peak identification.

10. Set the parameter for peak identification, annotation, and alignment as follows.
    - Retention index (RI) tolerance: 10.
    - Retention time (RT) tolerance: 0.5 min.
    - *m/z* tolerance: 0.5 Da.
    - EI similarity cutoff: 70%.
    - Identification score: 70%.
11. Normalize the height of detected and annotated peak using the height of internal standard peak.
12. Export the data matrix consisted of raw and normalized (*see* Subheading 4.1, **Note 5**) peak height of the detected and annotated peaks.
13. Filter out the detected peaks by excluding the peaks that were not of biological origin by referring to the blank (*see* Subheading 4.1, **Note 6**).
14. Filter out the detected peaks with a relative standard deviation of more than 30% based on the raw peak heights obtained from the QC (*see* Subheading 4.1, **Note 7**).
15. The data matrix with filtered peaks will be used for further multivariate analysis, such as PCA, OPLS-R, or OPLS-DA using SIMCA-P or R Studio.

3.2 Volatile Analysis

3.2.1 Sample Preparation

1. Green beans and roasted beans are first treated with liquid nitrogen and then ground into a fine powder.
2. Coffee brew made 8.25 g of roasted coffee grounds mixed with 150 mL of water at 93 ± 3 °C, ensuring that all the grounds are fully saturated. The brew is left to steep for 2 min before evaluation.

3.2.2 Extraction Green Bean Ground

1. A sample of 7 g of green bean ground was added with 1 μL of 0.01% 3-heptanone internal standard in methanol LC and placed in a 22 mL SPME vial.
2. Extraction was performed at 50 °C for 30 min using a 2 cm DVB/CAR/PDMS SPME fiber.
3. The fiber was then placed in the headspace for 2 min at 50 °C to absorb volatile compounds.

3.2.3 Extraction Roasted Bean Ground [28]

1. A sample of 5 g of roasted bean ground was added 10 μL of 0.01% 3-heptanone internal standard in methanol LC and placed in a 22 mL SPME vial.
2. Extraction was performed at 50 °C for 30 min using a 2 cm DVB/CAR/PDMS SPME fiber.
3. The fiber was then placed in the headspace for 2 min at 50 °C to absorb volatile compounds.

3.2.4 Extraction Coffee Brew

1. A sample of 0.85 g of coffee brew was added 1 μL of 0.01% 3-heptanone internal standard in methanol LC and placed in a 22 mL SPME vial.
2. Extraction was performed at 50 °C for 30 min using a 2 cm DVB/CAR/PDMS SPME fiber.
3. The fiber was then placed in the headspace for 2 min at 50 °C to absorb volatile compounds.

3.2.5 Gas Chromatography-Mass Spectrometry Analysis

1. Analysis was performed using a GC (Agilent 7890A, California, USA) equipped with a DB-Wax column (30 m × 250 μm × 0.25 μm) and an MS detector (Agilent 595975 XL EI/CI).
2. The GC injector was operated at 250 °C in splitless mode.
3. Helium gas flow rate set at 1 mL/min.
4. The oven temperature was set to 40 °C for 5 min, then increased by 3 °C/min until reaching 180 °C, followed by 10 °C/min until reaching 240 °C.
5. The ion source was maintained at 230 °C.
6. The MS Quad temperature at 150 °C.
7. The interface temperature at 250 °C.
8. Mass spectra were collected in the range of 29–550 amu.

3.2.6 Data Analysis

Analysis results were compared and matched with mass spectra patterns from the NIST 14 Library (National Institute of Standards and Technology). The LRI (linear retention index) of each identified compound was determined by injecting alkane standards (C_9–C_{29}). The relative concentration of the compounds (μg/kg) was calculated by comparing the peak area of each compound to that of the internal standard, 3-heptanone. The processing of identified compound data is carried out using PCA, PLS, or OPLS. The data can be analyzed using software such as XLSTAT, SIMCA-P, and R Studio.

4 Notes

4.1 Nonvolatile Analysis

1. It is often that the metal cone is stuck inside the tube during the preparation of green coffee beans powder. This occurs due to large debris of green (raw) coffee beans stuck between the metal cone and the tube, which could hinder the grinding process. To address this issue, it is suggested to submerge the tube in liquid nitrogen for a longer period.
2. The time between the completion of derivatization step and GC analysis might cause variance to the compound peak signal (peak height) [22]. It is recommended to inject the derivatized

samples into GC analysis within 24 h after the completion of derivatization reaction. Moreover, it is also advised to calibrate the GC-MS instrument prior to the derivatization step to ensure that the machine is functioning properly for analysis and prevent resource loss.

3. Depending on the instrument that is used for the analysis, the GC/MS analysis method for alkane mix might be different from sample. Method optimization is advised to obtain complete separation of each carbon standard peak and its respective retention time.
4. Depending on the instrument that is used for the analysis, the retention time of peak C9 until C40 could be detected (total 31 peaks) and retrieved from the instrument's software. The carbon numbers and their corresponding retention times should be organized in a table and saved as a text file using Notepad. Further, this file could be used as basis for retention index for peak annotation during MS-DIAL analysis. It is advised to refer to MS-DIAL manuals for details.
5. The term "raw peak height" refer to the peak before normalization using internal standard. The term "normalized peak height" refers to the peak height after raw peak height divided by the peak height of internal standard.
6. This step is conducted by comparing the height of detected peak between blank and the samples. The fivefold of blank raw peak height will be compared with the maximum height of the same peak obtained from the samples. If the fivefold of blank raw peak height is higher than the maximum raw peak height from the samples, the peak could be filtered out.
7. The relative standard deviation shows the differences of peak height within the QC, which contain all the samples. The smaller the number, the consistent the peaks existed in the sample with stable manner. It is expected that QC will have all the peaks that are contained in the samples. Thus, only QC are used for the relative standard deviation calculation.
8. The relative standard deviation is calculated based on standard deviation of the peak height divided by the average peak height times 100%. It can use the raw or normalized height of detected peak in the QC.

4.2 Volatile Analysis

1. The amount of internal standard added must be appropriate, with its intensity large enough to enable accurate calculations. Additionally, the internal standard should be used in a known quantity and be chemically similar to the analyte being measured but not present in the sample. The internal standard should consistently appear in the chromatogram, with a distinct peak that is clearly separated from the analyte peaks [29].

References

1. FAO (2015) FAO statistical pocketbook: coffee 2015. FAO
2. Buffo R (2018) Advances in research on coffee flavour compounds. In: Lashermes P (ed) Achieving sustainable cultivation of coffee. Burleigh Dodds Science Publishing, pp 291–309. https://doi.org/10.19103/as.2017.0022.15
3. Cordoba N, Pataquiva L, Osorio C, Moreno FLM, Ruiz RY (2019) Effect of grinding, extraction time and type of coffee on the physicochemical and flavour characteristics of cold brew coffee. Sci Rep 9(1):1–12. https://doi.org/10.1038/s41598-019-44886-w
4. Joët T, Dussert S (2018) Environmental and genetic effects on coffee seed biochemical composition and quality. In: Lashermes P (ed) Achieving sustainable cultivation of coffee. Burleigh Dodds Science Publishing, pp 49–61. https://doi.org/10.19103/as.2017.0022.15
5. Fisk ID, Kettle A, Hofmeister S, Virdie A, Kenny JS (2012) Discrimination of roast and ground coffee aroma. Flavour 1:1–9
6. Pereira GV d M, Neto DP d C, Júnior AIM, Vásquez ZS, Medeiros ABP, Vandenberghe LPS, Soccol CR (2019) Exploring the impacts of postharvest processing on the aroma formation of coffee beans – a review. Food Chem 272:441–452. https://doi.org/10.1016/j.foodchem.2018.08.061
7. Caporaso N, Whitworth MB, Cui C, Fisk ID (2018) Variability of single bean coffee volatile compounds of Arabica and robusta roasted coffees analysed by SPME-GC-MS. Food Res Int 108:628–640. https://doi.org/10.1016/j.foodres.2018.03.077
8. Putri SP, Irifune T, Yusianto, Fukusaki E (2019) GC/MS based metabolite profiling of Indonesian specialty coffee from different species and geographical origin. Metabolomics 15: 1–11
9. Król K, Gantner M, Tatarak A, Hallmann E (2020) The content of polyphenols in coffee beans as roasting, origin and storage effect. Eur Food Res Technol 246:33–39. https://doi.org/10.1007/s00217-019-03388-9
10. Amalia F, Aditiawati P, Yusianto, Putri SP, Fukusaki E (2021) Gas chromatography/mass spectrometry-based metabolite profiling of coffee beans obtained from different altitudes and origins with various postharvest processing. Metabolomics 17:1–16
11. Herawati D, Davin C, Yuliana ND, Yulianti Y (2024) Effect of storage duration on phenolics stability in ready-to-drink coffee beverage. Coffee Sci 19. https://doi.org/10.25186/.v19i.2194
12. Yulianti Y, Adawiyah DR, Herawati D, Indrasti D, Andarwulan N (2024) Identification of antioxidant and flavour marker compounds in Kalosi-Enrekang Arabica brewed coffee processed using different postharvest treatment methods. J Food Sci Technol 61: 1165–1179. https://doi.org/10.1007/s13197-024-05948-8
13. Angeloni S, Mustafa AM, Abouelenein D, Alessandroni L, Acquaticci L, Nzekoue FK et al (2021) Characterization of the aroma profile and main key odorants of espresso coffee. Molecules 26(13):3856
14. Yu JM, Chu M, Park H, Park J, Lee KG (2021) Analysis of volatile compounds in coffee prepared by various brewing and roasting methods. Foods 10(6):1347
15. Farag MA, Zayed A, Sallam IE, Abdelwareth A, Wessjohann LA (2022) Metabolomics-based approach for coffee beverage improvement in the context of processing, brewing methods, and quality attributes. Foods 11(6):864
16. Sittipod S, Schwartz E, Paravisini L, Peterson DG (2019) Identification of flavor modulating compounds that positively impact coffee quality. Food Chem 301:125250
17. Rocchetti G, Braceschi GP, Odello L, Bertuzzi T, Trevisan M, Lucini L (2020) Identification of markers of sensory quality in ground coffee: an untargeted metabolomics approach. Metabolomics 16:1–12
18. Amalia F, Irifune T, Takegami T, Yusianto, Sumirat U, Putri SP, Fukusaki E (2023) Identification of potential quality markers in Indonesia's Arabica specialty coffee using GC/MS-based metabolomics approach. Metabolomics 19(11):90
19. Yulianti Y, Adawiyah DR, Herawati D, Indrasti D, Andarwulan N (2023) Detection of markers in green beans and roasted beans of Kalosi-Enrekang Arabica coffee with different postharvest processing using LC-MS/MS. Int J Food Sci 2023. https://doi.org/10.1155/2023/6696808
20. Pua A, Goh RMV, Huang Y, Tang VCY, Ee KH, Cornuz M, Liu SQ, Lassabliere B, Yu B (2022) Recent advances in analytical strategies for coffee volatile studies: opportunities and challenges. Food Chem. https://doi.org/10.1016/j.foodchem.2022.132971
21. Orata F (2012) Derivatization reactions and reagents for gas chromatography analysis. In: Mohd MA (ed) Advanced gas

chromatography-Progress in agricultural, biomedical and industrial applications. InTech. https://doi.org/10.5772/33098

22. Miyagawa H, Bamba T (2019) Comparison of sequential derivatization with concurrent methods for GC/MS-based metabolomics. J Biosci Bioeng 127(2):160–168
23. Płotka-Wasylka J, Szczepańska N, de la Guardia M, Namieśnik J (2015) Miniaturized solid-phase extraction techniques. TrAC Trends Anal Chem 73:19–38. https://doi.org/10.1016/j.trac.2015.04.026
24. Aditiawati P, Astuti DI, Kriswantoro JA, Khanza SM, Kamarisima, Irifune T, Amalia F, Fukusaki E, Putri SP (2020) GC/MS-based metabolic profiling for the evaluation of solid state fermentation to improve quality of Arabica coffee beans. Metabolomics 16:1–14
25. Lai Z, Tsugawa H, Wohlgemuth G, Mehta S, Mueller M, Zheng Y, Ogiwara A, Meissen J, Showalter M, Takeuchi K, Kind T, Beal P, Arita M, Fiehn O (2018) Identifying metabolites by integrating metabolome databases with mass spectrometry cheminformatics. Nat Methods. https://doi.org/10.1038/nmeth.4512
26. Tsugawa H, Cajka T, Kind T, Ma Y, Higgins B, Ikeda K, Kanazawa M, Vandergheynst J, Fiehn O, Arita M (2015) MS-DIAL: data-independent MS/MS deconvolution for comprehensive metabolome analysis. Nat Methods. https://doi.org/10.1038/nmeth.3393
27. Tsugawa H, Nakabayashi R, Mori T, Yamada Y, Takahashi M, Rai A, Sugiyama R, Yamamoto H, Nakaya T, Yamazaki M, Kooke R, Bac-Molenaar JA, Oztolan-Erol N, Keurentjes JJB, Arita M, Saito K (2019) A cheminformatics approach to characterize metabolomes in stable-isotope-labeled organisms. Nat Methods. https://doi.org/10.1038/s41592-019-0358-2
28. Jumhawan U, Putri SP, Yusianto, Marwani E, Bamba T, Fukusaki E (2013) Selection of discriminant markers for authentication of Asian palm civet coffee (Kopi Luwak): a metabolomics approach. J Agric Food Chem 61(33): 7994–8001
29. Ouyang G (2012) Calibration. In: Handbook of solid phase microextraction. Elsevier Inc, pp 167–199. https://doi.org/10.1016/B978-0-12-416017-0.00006-1

Chapter 6

GC-MS-Based Metabolite Profiling Method for Durian

Pinnapat Pinsorn and Supaart Sirikantaramas

Abstract

In this chapter, we describe volatile and nonvolatile metabolite identification methods in durian pulp using gas chromatography-mass spectrometry (GC-MS). Special focus is given to sulfur-containing volatile compounds, such as methanethiol, ethanethiol, and hydrogen sulfide, as well as key metabolites that contribute to the characteristic aroma of durian. In addition, we detail the methodologies for metabolite extraction, chemical derivatization, and identification of primary metabolites, including sugars, amino acids, and organic acids. Moreover, we provide a collection of durian samples and highlight the importance of defining maturity and ripening stages, as these factors influence metabolite composition. The methods described in this chapter could serve as a model for studying metabolite profiling in other fruits, with potential applications in food science and agricultural research.

Key words Durian, Fruit, Pulp, GC-MS, Derivatization, Headspace, Sulfur, Volatile

1 Introduction

Durian is an economically significant tropical fruit native to Southeast Asia. It is commercially cultivated in several countries, including Thailand, Malaysia, and Indonesia. Durian displays a uniquely pungent flavor due to the production of sulfur-containing aromas during ripening [1]. Its high carbohydrate, sugar, and fatty acid content make it an energy-rich food [2]. In addition, durian contains bioactive compounds, such as carotenoids and antioxidants, thereby offering health benefits. Furthermore, durian pulp is a rich source of kokumi substances, which contribute to an intense taste and mouthfeel, including gamma-glutamylcysteine and glutathione [1, 3].

In our previous studies on durian fruit, we used several high-throughput techniques, such as capillary electrophoresis–mass spectrometry, liquid chromatography-mass spectrometry, and gas chromatography-mass spectrometry (GC-MS), to access metabolites in durian pulp at different fruit stages [1, 4, 5]. Among various analytical techniques, GC-MS has proven to be a powerful tool for

Sastia Prama Putri (ed.), *Gas Chromatography-Mass Spectrometry: Based Metabolomics in Food Science*, Methods and Protocols in Food Science, https://doi.org/10.1007/978-1-0716-5108-7_6,

both volatile and nonvolatile compound identification and quantification in durian. GC-MS allows for the comprehensive profiling of primary metabolites (such as sugars, amino acids, and organic acids) as well as volatile compounds responsible for the distinctive odor and taste of durian [4, 5]. Concerning nonvolatile compounds, chemical derivatization is essential and must be performed prior to GC-MS analysis. For the determination of volatile compounds (e.g., major odor-active compounds: methanethiol, ethanethiol, and hydrogen sulfide), GC-MS-coupled headspace analysis is a commonly used technique [6].

In this chapter, we provide an in-depth analysis of durian metabolite composition using GC-MS, presenting the sample preparation, extraction, and data analysis methods in detail. In addition, we provide a detailed description of durian sample collection and clarify that the fruit ripening stage is a crucial aspect of studying metabolites. The results we present in this chapter contribute to a broader understanding of the biochemical properties of durian, and the methods applied to studying durian could serve as a model for investigating other fruits with a similar approach.

2 Materials

2.1 Plant Material

1. Durian pulp (fresh or freeze-dried).

2.2 Reagents and Chemicals

1. Liquid nitrogen.
2. Thiophene (internal standard for volatile analysis).
3. Methanol (LC-grade).
4. Chloroform (LC-grade).
5. Ultrapure water/LC-grade water.
6. $^{13}C_6$-Sorbitol (internal standard for the analysis with chemical derivatization).
7. n-Alkane standard mix (for retention index calculations).
8. Argon or nitrogen gas.
9. Methoxyamine hydrochloride.
10. N-Methyl-N-(trimethylsilyl)trifluoroacetamide (MSTFA, silylation derivatization reagent, CAS no. 24589-78-4).

2.3 Equipment

1. 50-mL tube.
2. 1.5-mL/2.0-mL microtube (Eppendorf tube is recommended).
3. GC glass vials with crimp or screw caps and chemically inert septa.
4. Crimper.

5. Lyophilizer or freeze-dryer.
6. Mixer mill/grinding machine.
7. Balance.
8. Texture analyzer.
9. Microwave oven.
10. Pipette.
11. Vortex mixer.
12. Microtube centrifuge.
13. Shaker incubator.
14. Speed vacuum concentrator.
15. Gas chromatography-mass spectrometry system.
16. GC column: HP-INNOWax column (part no. 19091N-33; length: 30 m; inner diameter: 0.25 mm; film thickness: 0.25 μm).
17. GC column: 5% phenyl–95% dimethylpolysiloxane fused silica capillary column (length: 30 m; inner diameter: 0.25 mm; film thickness: 0.25 μm).

3 Methods

To simplify the process overview starting from sample collection, extraction, and derivatization, we provide a flowchart outlining the procedures for durian collection and volatile and nonvolatile compound analysis using GC-MS (Fig. 1).

3.1 Sample Collection and Freeze-Drying

Durian fruit are harvested from trees at a specific stage. As durian is a climacteric fruit, we harvested it based on its maturity stage, determined by the percentage of dry matter according to the Thai Agricultural Standard (TAS 3-2024) (*see* **Notes 1–7**) [7], to study postharvest ripening. For example, we harvested the Chanee and Monthong cultivars approximately 90 and 110 days after anthesis, respectively, when the dry matter content exceeded 30% and 32%, respectively [1]. We measured fruit firmness to determine the ripening stage using a texture analyzer. Table 1 presents the details of each durian ripening stage.

1. Peel the fruits when they reach the appropriate stage and collect the pulp.
2. Transfer the fresh pulp into a 50-mL tube, which we then either immediately dip in liquid nitrogen to stop biological processes or use directly for volatile compound analysis (*see* **Note 8**).

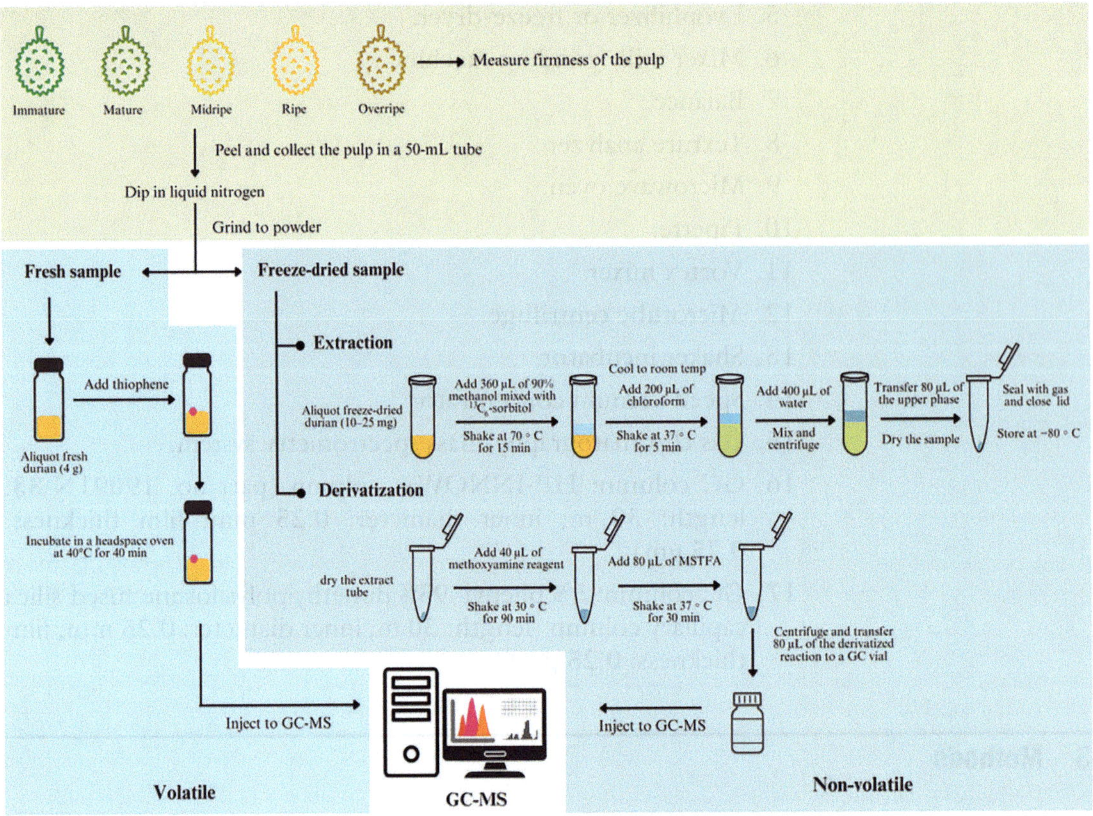

Fig. 1 Flowchart of the procedures for durian collection and analysis of volatile and nonvolatile compounds using GC-MS

3. Grind fresh durian pulp using a Mixer Mill by placing the frozen pulp in a stainless-steel chamber with a 15-mm grinding ball (*see* **Note 9**).
4. For volatile compound analysis using static headspace GC-MS, insert 4 g of frozen ground sample into a 20-mL GC glass vial. Add 200 μL of 10-ppm thiophene as an internal standard. Seal the vial immediately and place it in the headspace oven for analysis (*see* **Note 10**).

3.2 Metabolite Extraction for Primary Metabolite Profile

1. Prepare fresh 90% methanol-water (v/v) mixed with $^{13}C_6$-sorbitol (0.02 mg/mL).
2. Prepare fresh chloroform mixed with soluble internal standards (e.g., 0.25 mg/mL nonadecanoic acid methyl ester) for non-polar metabolite analysis, if needed.
3. Aliquot freeze-dried durian samples (10–25 mg) into 2.0-mL microtubes.
4. Add 360 μL of methanol mixed with $^{13}C_6$-sorbitol.

Table 1
Ripening stages of commercial durian cultivars in Thailand

Durian cultivar	Maturity stage (DAA)[a]	Dry matter (%)[b]	Ripening stage	Days after harvesting[c]	Firmness (*N*)[c]
Chanee	<90	<30	Immature	–	38.67 ± 2.23
	90	30	Mature (Unripe)	0	52.14 ± 1.97
			Midripe	2	3.14 ± 0.84
			Ripe	4	1.70 ± 0.45
			Overripe	6	1.13 ± 0.41
Monthong	<110	<32	Immature	–	43.2 ± 0.92
	110	32	Mature (Unripe)	0	48.13 ± 1.08
			Midripe	3	3.85 ± 1.71
			Ripe	5	2.02 ± 0.25
			Overripe	8	1.39 ± 0.45

[a]For commercial durian cultivars in Thailand, the fruits are harvested from trees at the maturity stage, which varies by cultivar. At this stage, the fruits are mature and able to naturally produce ethylene and ripen after harvesting

[b]The dry matter percentage standard value for Chanee and Monthong was measured and calculated using the formula in Eq. 1 (*see* **Notes 1–7**). If the dry matter of a harvested fruit remains below the standard value (30% for Chanee and 32% for Monthong), the fruit is classified as immature or of low quality

[c]The days after harvesting refer to the period during which the fruit is left to ripen naturally at room temperature (approximately 30 °C). Each recorded ripening stage is approximate and may vary by ±1 to 2 days. Therefore, firmness is necessary for stage confirmation using a texture analyzer

5. Mix thoroughly using a vortex mixer for 15 s and continue shaking at 70 °C for 15 min.
6. Cool the samples to room temperature (25 °C).
7. Add 200 μL chloroform mixed with the internal standard (if needed).
8. Mix the samples thoroughly for 15 s and continue shaking at 37 °C for 5 min.
9. Add 400 μL LC-grade water into the tubes.
10. Mix the samples thoroughly using a vortex mixer for 15 s.
11. Centrifuge the tubes at 14,000 rpm for 10 min to separate the liquid from the debris.
12. Transfer 80 μL of the supernatant upper phase containing methanol and water (polar phase) to a new 1.5-mL tube. Aliquot for backup (e.g., 100 μL and 200 μL in separate 1.5-mL tubes).
13. Dry the samples using a seed vacuum until completely dry (usually for 3–4 h).
14. Seal the dry samples under argon or nitrogen gas, close the lid, and store the samples at −80 °C for further analysis.

3.3 Derivatization

Derivatization reactions should be performed to modify the nonvolatile compounds. We describe below a previously published protocol [8]. Before starting derivatization, we took the samples out of the freezer and dried them using a speed vacuum concentrator for 30 min to remove any moisture that had accumulated after we took the samples out. Prepare a fresh methoxyamine reagent by dissolving methoxyamine hydrochloride in pyridine (40 mg/mL). Add 40 μL of methoxyamine reagent to the dried extract, then prepare one additional empty 1.5-mL tube as a "blank."

1. Mix using a vortex mixer, then shake continuously at 30 °C for 90 min, tapping the tubes every 30 min.
2. Add 80 μL of MSTFA to the samples, vortex them thoroughly, and incubate the tubes at 37 °C for 30 min.
3. Centrifuge the samples at 14,000 rpm for 5 min to precipitate any undissolved particles (optional).
4. Transfer 80 μL of the derivatized reactions into GC vials, close the cap immediately, and keep the GC vials at room temperature in a GC injector.
5. Prepare a standard mixture of n-alkanes in pyridine in a GC vial.

3.4 GC-MS Conditions

Here, to profile the volatiles, especially sulfur-containing types, we used a static headspace coupled with GC-MS. We used a Triple Quadrupole GC-MS/MS from Agilent Technologies, coupled with an Agilent 7697A static headspace sampler and Electron Impact Ionization (EI) system. Equivalent specifications can be applied. Tables 2 and 3 summarize the headspace and GC conditions.

In this chapter, we describe the use of an HP-INNOWax column for identifying small sulfur-containing volatile organic compounds. However, other columns, such as a 5% phenyl–95% dimethylpolysiloxane fused silica capillary column (length: 30 m;

Table 2
Headspace parameters

Parameter	Condition	
Oven	Oven temperature	40 °C
	Loop temperature	55 °C
	Transfer line temperature	125 °C
	Equilibration time	40 min
Time	Injection time	30 s
	Loop size/injection volume	1 mL
	GC cycle time	70 min

Table 3
GC parameters for volatile analysis

Parameter	Condition	
Injection	Inlet mode	Headspace
Column	HP-INNOWax column Helium carrier flow rate	 1 mL/min (constant flow)
Oven (Initial)	Initial temperature Hold time	35 °C 5 min
Oven (Ramp1)	1st ramp temperature Ramp rate	180 °C 5 °C/min
Oven (Ramp2)	2nd ramp temperature Ramp rate	250 °C 40 °C/min
Oven (hold)	Temperature Hold time	250 °C 5 min
Mass spec detector	Solvent delay MS scan range Transfer line temperature EI source temperature EI voltage	0 min 29–400 m/z 250 °C 230 °C 70 eV

inner diameter: 0.25 mm; film thickness: 0.25 μm) or equivalent, could be used for general purposes or selected based on the target compounds to be analyzed.

To measure the primary metabolites, we used an Agilent 6890 N gas chromatograph with a split/splitless injector. Equivalent specifications could be applied. Table 4 contains the GC conditions.

3.5 Data Analysis

Calculate retention indices based on the n-alkane standard mix for accuracy and consistency. Annotate the metabolites based on the known standard compounds, the Golm Metabolome Database (http://gmd.mpimp-golm.mpg.de/), National Institute of Standards and Technology (NIST) Mass Spectral Library, or Wiley MS libraries. Perform statistical analyses using an appropriate method.

4 Notes

1. (Optional) The percentage of dry matter of durian pulp is measured as follows:
2. Peel the durian fruit to be tested and cut it crosswise at the central part of the aril. Next, slice it into approximately 2–2.5-cm thick pieces.

Table 4
GC parameters for the primary metabolites

Parameter	Condition	
Injection	Inlet mode	Splitless Split mode (ratio 1:30) (*see* **Note 11**)
	Injection temperature	230–250 °C
	Injection volume	1 μL
Column	5% phenyl–95% dimethylpolysiloxane fused silica capillary column (preferred)	
	Helium carrier flow rate	0.6 mL/min (constant flow)
Oven (initial)	Initial temperature	70 °C
	Hold time	1 min
Oven (Ramp1)	1st ramp temperature	350 °C
	Ramp rate	9 °C/min
Oven (hold)	Temperature	350 °C
	Hold time	5 min
Mass spec detector	MS scan range	70–600 m/z
	Transfer line temperature	250 °C
	EI source temperature	250 °C
	EI voltage	70 eV

3. Chop the durian pulp into small pieces.
4. Place an empty plate and record its fresh weight.
5. Place an empty paper plate in a microwave oven and dry it using low heat for 2–3 min and weigh the dried paper plate and record its weight.
6. Weigh 10 g of the chopped durian pulp and place it on the pre-weighed paper plate.
7. Put the durian pulp and the paper plate into the microwave oven. Dry using low to medium heat for consistent intervals of 2–3 min and weigh after each interval. Repeat this process until the weight of the durian pulp remains constant. Calculate the dry matter percentage of the durian pulp using the following formula (Eq. 1):

$$\%\textit{dry}\text{ matter} = (\textit{dry}\text{ weight (g)})/(\text{Fresh weight (g)}) \times 100 \quad (1)$$

8. Avoid direct contact between the liquid nitrogen and the samples, ensuring that the lid was properly closed to prevent leakage, as Falcon tubes might explode if liquid nitrogen is trapped inside.

9. Frozen durian samples were rather difficult to break. Therefore, grinding at 30 Hz for two rounds of 2 min each was more effective. The stainless-steel chambers and grinding balls should be precooled in liquid nitrogen before placing the durian pulp inside. Between rounds, the chamber should be dipped in liquid nitrogen before proceeding to the second round.
10. Prepare a 10-ppm thiophene solution in LC-grade water. After weighing 4 g of durian pulp, pipette 200 μL of the 10-ppm thiophene solution and immediately seal the cap.
11. The split ratio can be adjusted to keep the metabolite abundance within the quantifiable range.

References

1. Pinsorn P, Oikawa A, Watanabe M, Sasaki R, Ngamchuachit P, Hoefgen R, Saito K, Sirikantaramas S (2018) Metabolic variation in the pulps of two durian cultivars: unraveling the metabolites that contribute to the flavor. Food Chem 268:118–125. https://doi.org/10.1016/j.foodchem.2018.06.066
2. Khaksar G, Kasemcholathan S, Sirikantaramas S (2024) Durian (Durio zibethinus L.): nutritional composition, pharmacological implications, value-added products, and omics-based investigations. Horticulturae 10:342. https://doi.org/10.3390/horticulturae10040342
3. Singcha P, Khaksar G, Sirijan M, Sirikantaramas S (2024) Durian (*Durio zibethinus* L.) fruit: a superior dietary source of natural glutathione and γ-glutamylcysteine. J Food Compos Anal 127:105975. https://doi.org/10.1016/j.jfca.2024.105975
4. Pinsorn P, Sanachai K, Rungrotmongkol T, Hoefgen R, Watanabe M, Brueckner F, Nakabayashi R, Mori T, Oikawa A, Sasaki R, Sucharitakul J, Saito K, Sirikantaramas S (2023) Sulfur metabolism in durian pulps: factors contributing to the production of volatile sulfur compounds during fruit ripening. Postharvest Biol Technol 206:112533. https://doi.org/10.1016/j.postharvbio.2023.112533
5. Sangpong L, Khaksar G, Pinsorn P, Oikawa A, Sasaki R, Erban A, Watanabe M, Wangpaiboon K, Tohge T, Kopka J, Hoefgen R, Saito K, Sirikantaramas S (2021) Assessing dynamic changes of taste-related primary metabolism during ripening of durian pulp using metabolomic and transcriptomic analyses. Front Plant Sci 12:687799. https://doi.org/10.3389/fpls.2021.687799
6. Li JX, Schieberle P, Steinhaus M (2012) Characterization of the major odor-active compounds in Thai durian (Durio zibethinus L.'Monthong') by aroma extract dilution analysis and headspace gas chromatography–olfactometry. J Agric Food Chem 60:11253–11262
7. Office of Agricultural Commodity and Food Standards (ACFS) (2024) มาตรฐานสินค้าเกษตร: ทุเรียน (มกษ. 3–2567). Ministry of Agriculture and Cooperatives, Thailand.
8. Erban A, Martinez-Seidel F, Rajarathinam Y, Dethloff F, Orf I, Fehrle I, Alpers J, Beine-Golovchuk O, Kopka J (2020) Multiplexed profiling and data processing methods to identify temperature-regulated primary metabolites using gas chromatography coupled to mass spectrometry. In: Hincha D, Zuther E (eds) Plant cold acclimation. Methods in molecular biology, vol 2156. Humana, New York, pp 203–239. https://doi.org/10.1007/978-1-0716-0660-5_15

Chapter 7

GC-MS-Based Metabolite Profiling Methods for Horticultural Crops

Yimeng Zhao and Farhana R. Pinu

Abstract

GC-MS is one of the most widely used analytical instruments for metabolite profiling of different biological samples. There are many methodologies involving different types of metabolite extraction and sample preparation strategies available in published literature. For horticultural crops, GC-MS has been used to analyze both volatile and semi/nonvolatile compounds. To analyze volatile compounds, GC-MS is undeniably the most suitable analytical platform as it allows direct detection of volatile compounds within the samples. However, semi/nonvolatile metabolites can also be analyzed in horticultural crops using GC-MS and many efforts have been made to develop and optimize different derivatization protocols. This book chapter covers a detailed description of different methodologies available for the analysis of both volatile and nonvolatile metabolites in different fruit and horticultural products using both targeted and untargeted approaches by GC-MS. In addition to methodologies, key considerations that need to be considered while performing GC-MS-based metabolite profiling will also be outlined.

Key words Volatiles, Nonvolatiles, Sample preparation, Extraction, Mass spectrometry, Detection, Identification

1 Introduction

Since the invention and development in the 1950s, GC-MS has become one of the most preferred analytical platforms owing to its versatility, sensitivity, robustness, sensitivity, and excellent separation capabilities. Although initially, GC-MS was mainly used for targeted analysis of a few volatile compounds present in a mixture, the technology has advanced enough to be able to analyze hundreds of compounds in any sample simultaneously. Therefore, metabolite profiling using GC-MS became a popular approach and is now widely used to analyze various biological samples [1]. Metabolite profiling involves the separation, detection, and identification of as many metabolites as possible in a given sample [2, 3]. Since the introduction of metabolomics approaches, many food and

Sastia Prama Putri (ed.), *Gas Chromatography-Mass Spectrometry: Based Metabolomics in Food Science*, Methods and Protocols in Food Science, https://doi.org/10.1007/978-1-0716-5108-7_7,

horticultural scientists have opted toward using metabolite profiling to generate novel insights on biological systems [4–8].

GC-MS-based metabolite profiling has been used for the analysis of both primary and secondary metabolites. Particularly, it is relatively easy to analyze volatile compounds present in any sample simply by extracting them using suitable solvents or directly from the headspace of the samples. However, nonvolatile compounds need to be derivatized to make them volatile enough to be analyzed by GC-MS [1]. Significant work has been conducted so far to develop and optimize parameters to extract different classes of metabolites, which depend on the type of sample. Similarly, various chemical derivatization protocols are also available.

The application of GC-MS to analyze different horticultural crops benefited from generating detailed metabolite fingerprints [9], allowing the differentiation between crops based on their overall metabolite composition. Using GC-MS data, it was possible to elucidate metabolic pathways in different plants by examining the overall pattern of primary and secondary metabolites [10]. For example, GC-MS can efficiently identify and quantify a wide range of primary metabolites (e.g., free amino acids, acids, fatty acids, and sugars), which play a crucial role in the nutritional value and physiological responses of plants [11, 12]. The successful identification of starch in potatoes, fructose in fruits, glutamic acid in tomatoes [13], oleic acid in olives [14], linoleic acid in nuts and seeds [15], acids in lemons [16], and raspberry [17] using GC-MS has highlighted its importance in understanding plant physiology and complex metabolic networks. Additionally, GC-MS is frequently employed in profiling secondary metabolites to provide a more comprehensive understanding of the plant metabolome [18, 19]. A variety of compounds, including flavonoids, alkaloids, and terpenoids, were the main secondary metabolites analyzed using GC-MS-based techniques. Studies have used GC-MS to analyze complex profiles of mixtures of primary and secondary metabolites from various horticultural species, providing important insights into their biochemical pathways and ecological interactions [20, 21].

This book chapter aims to provide a detailed description of popular GC-MS methodologies used for the metabolite profiling of different horticultural crops, along with considerations that need to be accounted for while preparing and analyzing the samples. In addition, the advancements and drawbacks related to data processing and analysis will also be covered.

1.1 Overview of the Use of GC-MS in Horticultural Science

GC-MS-based technologies make use of a combination of targeted and nontargeted methods to facilitate the detection of specific classes of compounds. However, metabolite profiling approaches should be conducted in an untargeted manner, allowing the detection and identification of as many metabolites as possible. This way, it is possible to discover novel metabolites and related metabolic pathways. With its powerful qualitative and quantitative analytical capabilities, GC-MS has become an important supporting technology in characterizing ingredients, assessing crop and food quality and safety, and understanding plant responses to environmental factors. It has been extensively utilized in analyses of VOCs, characterizations of sugars and phenols that affect aroma, and the tracking of dynamics of key metabolites during maturation and stress [22, 23]. Furthermore, the utilization of GC-MS in conjunction with non-compositional or functional targeting studies facilitates the detection of pesticide residues and contaminants, thereby ensuring food product safety [24]. With the continuous progress of analytical technology, the application of GC-MS in future horticultural crop research will develop in the direction of high-throughput, intelligent integration and become an essential tool for quality control, safety monitoring, and sustainable agriculture. Table 1 summarizes the key applications of GC-MS-based methodologies from published literature to highlight the importance of this technology in different aspects of horticultural science.

Table 1 summarizes key GC-MS applications from existing literature, demonstrating the broad utility of this platform in horticultural science.

2 Materials

2.1 General Sample Handling

1. Liquid nitrogen.
2. Extremely low temperature storage (−80 °C).

2.2 Solvent-Free Volatile Profiling (HS-SPME, Apple)

1. Sodium chloride (NaCl, 30% w/w).
2. Internal standards: Cyclohexanone (2.025 μg/sample).
3. SPME fiber: 1 cm, 50/30 μm DVB/CAR/PDMS.

2.3 Solvent Extraction of Volatiles (Balsamic Vinegar)

1. Ether (extraction solvent).
2. Internal standard: 12-bromo-1-dodecanol (1 mM, 5 μL/sample).
3. Nitrogen gas (N_2) for concentration.

2.4 Trimethylsilyl (TMS) Derivatization

1. Derivatization reagents: methoxyamine hydrochloride solution in pyridine (2 mg/100 mL) and N-methyl-N (trimethylsilyl)-trifluoroamide (MSTFA).
2. Internal standard: D-ribitol.

2.5 Methyl Chloroformate (MCF) Derivatization

1. Internal standard: L-alanine-2,3,3,3-d_4 (10 mM).
2. 1 M NaOH, methanol, pyridine, sodium bicarbonate solution (50 mM), and anhydrous sodium sulfate.

Table 1
GC-MS applications in horticultural science

Sample	Metabolites analyzed	Extraction	Derivatization	References
Tomato	Biogenic VOCs	Dynamic headspace sampling	N/A	[25]
Pea	VOCs, saponins	Headspace solid phase microextraction	Phosphate buffer (pH 5.5), solvent extraction (80/20 MeOH/H_2O)	[26]
Coconut	VOCs, amino acids, sugars, organic acids, lipids, vitamins, flavonoids	Solvent extraction (80/20 MeOH/H_2O)	N/A	[27]
Walnut	Naphthoquinones, fatty acid esters, flavonoids, tannins, alkaloids, and phenolic acids, organic acids	Solvent extraction (96/4 EtOH/H_2O)	N/A	[28]
Navel orange	VOCs	Headspace solid phase microextraction	N/A	[29]
Strawberry	VOCs	Headspace solid phase microextraction	N/A	[30]
Mulberry	VOCs, flavonoids, phenolic acids, saponins	Headspace solid phase microextraction	N/A	[31]
Mulberry	VOCs, amino acids, organic acids, sugars, phenolics, vitamins, fatty acids, cyanidins	Solvent extraction (MeOH and hexane)	Methoximation (MeOX) + Silylation (BSTFA and TMCS)	[32]
Strawberry	VOCs	Headspace solid phase microextraction	N/A	[33]
Apple	VOCs	Solvent extraction (MeOH, chloroform: Methanol:Water)	Methoximation (MeOX) + Silylation (BSTFA)	[34]
Cranberry	Benzoic acids, phenolic acids	Liquid-liquid extraction (ether and ethyl acetate)	Silylation (BSTFA and TMCS)	[35]
Eggplant leaves	Free fatty acids, triacylglycerols, fatty acid methyl esters	Solvent extraction (chloroform/ methanol 2:1)	Trimethylsilylation (BSTFA + TMCS), tert-Butyldimethylsilylation (MTBSTFA + TBDMC), methylation (TMSD, BF_3/MeOH)	[36]

(continued)

Table 1
(continued)

Sample	Metabolites analyzed	Extraction	Derivatization	References
Carrots, tomatoes, lettuce, onions	Hormones, pharmaceuticals, industrial pollutants, flame retardants	Solvent extraction with ultrasonic assistance (EtAc/MeOH 9:1+ NH_4OH and formic acid)	Trimethylsilylation (BSTFA /TMCS 99:1)	[37]
Bean sprouts	Pesticide growth regulators	Liquid-liquid extraction with sonication & salting-out (acetonitrile & NaCl), solid phase extraction	Methylation (TMSD, BF_3/MeOH)	[38]
Pineapple	VOCs	Headspace solid phase microextraction	N/A	[39]
Tamarillo	Organic acids	Solvent extraction (MCF)	MCF, chloroform, sodium acetate buffer (pH 5.3. 1.5 M)	[40]
Mango, pineapple, jackfruit, baobab, tamarind	Organic acids, phenolics, sugars, VOCs	Solvent extraction (85/15 MeOH/ H2O), hydrochloric acid (0.5 mL of 0.75 M), diethyl ether	Silylation (TMSCN)	[41]
Pumpkin	Amino acids, fatty acids, sterols, carotenoids, polyphenols	Solvent extraction (70/30 EOH/H2O), petroleum ether, methylene chloride, ethyl acetate, n-butanol	98/2 MeOH/H_2SO_4	[42]

BSTFA N,O-bis(trimethylsilyl)trifluoroacetamide, *TMCS* trimethylchlorosilane, *MTBSTFA* *N-tert*-butyldimethylsilyl-*N*-methyltrifluoroacetamide, *TBDMC* tert-butyldimethylchlorosilane; TMSD, (trimethylsilyl)diazomethane, *MSTFA*, *N*-Methyl-*N*-(trimethylsilyl)trifluoroacetamide, *MeOH* methanol, *EtAc* ethyl acetate, *MCF* methyl chloroformate, *TMSCN* trimethylsilyl cyanide

2.6 Fatty Acid Esterification

1. Methanol, chloroform, acetyl chloride, 6% K_2CO_3 solution.
2. Internal standards: C23:0 (0.48 mg in 50 mL chloroform) and C19:0 solution in methanol:toluene 4:1 (v/v).

2.7 Sample Preparation and Storge

Sample preparation is one of the major critical steps in the overall GC-MS framework. Before preparing any samples, appropriate care and attention should be paid to understand the nature of analytes and the sample matrix. It is important to note that proper handling prevents the deterioration, contamination, or loss of the target analyte, thus avoiding compromising accuracy and reproducibility [43]. In horticultural research, many factors, including plant material selections, collection timing, and post-collection processing, all play a critical role in ensuring the integrity and reliability of analytical results [44]. The variability in chemical properties and composition of bioactive components of different plant materials (e.g., powders, juices, purees, essential oils) and their different parts (e.g., roots, stems, leaves, flowers, and fruits) has been thoroughly explored in many studies [45, 46]. Additionally, ecological variations, climate changes, and environmental factors have been shown to result in considerable regulations on the physiological dynamics and composition of plant metabolites [47, 48]. Therefore, researchers should take these factors into consideration when developing sample collection protocols to ensure the stability of the metabolites to be analyzed, thus ensuring the reliability and reproducibility of the collected data. Following a comprehensive evaluation of the crop sample's characteristics, post-collection handling is the next step. These handlings require considerations of storage conditions, effects of plant matrix, and the role of cleaning treatments to minimize interference [1, 43]. For volatile metabolite profiling, purging the sample containers or vials using an inert gas (e.g., nitrogen or argon) is a common practice to create an oxygen-free condition. In case of nonvolatile compounds, samples are often quenched with liquid nitrogen followed by a grinding or homogenization if samples are tissues (solid) so that active enzymes are inactivated, thus stopping further metabolic activities [44]. One must also avoid cross-contaminating samples, and precautions should be taken, mainly by preparing blank samples to account for any source of contamination that may arise from reagents, glasswares, and even from pipettes used during the extraction and derivatization steps.

After preparing the samples, they are often stored prior to the analysis, and the sample storage conditions directly influence analyte stability, integrity, and detection by GC-MS analysis. Factors, including temperature, humidity, light exposure, and container selection, have been demonstrated to result in varying levels of analyte degradation, oxidation, metabolite loss, or chemical transformation. Many metabolites in crops, such as terpenes and aromatic molecules, exhibit high levels of volatility and thus potentially alter the chemical profile of the samples [49, 50]. Furthermore, oxidation and degradation cause significant risks during storage as well, particularly for phenolic compounds, flavonoids, some amino acids, and fatty acids [51, 52]. Meanwhile, enzymes such as lipase and peroxidase in fresh plant tissues also contribute to degrading lipids and phenolic compounds, leading to compositional changes [53]. These generations of by-products that interfere with the target analyte are particularly problematic for metabolite profiling. To prevent these undesirable conditions, most horticultural research nowadays involves the use of ultra-freezing conditions (i.e., temperatures of −20 °C or − 80 °C) to store samples in airtight containers and sometimes in dark containers if metabolites are sensitive to lights. In addition to subjecting samples directly in freezing facilities, dry ice and liquid nitrogen are often used to accelerate the freezing process to form smaller ice crystals that are more effective in preserving the integrity of plant tissue [54, 55]. This strategy has been confirmed to be effective in retaining such sensitive nutrients (vitamin C, carotenoids, polyphenols) and VOCs in studies based on a large number of fruits and vegetables, such as blueberries, raspberries, peaches, pineapples, plums, cabbage, carrots, and others [56–60].

3 Methods

GC-MS has been used to analyze both primary and secondary metabolites in different horticultural crops, as shown in Table 1. Moreover, much research has been conducted to develop and optimize methodologies to determine volatile and nonvolatile compounds. However, there is still no universal extraction and analytical platform that allows the simultaneous determination of both volatile and nonvolatile components present in a sample by GC-MS [61]. Therefore, GC-MS-based metabolite profiling used in horticultural crops can be regarded as volatile and nonvolatile fingerprinting. These types of analyses are usually semi-quantitative, although methods for absolute quantification of metabolites have been published already [62, 63].

Figure 1 represents a general overview of the GC-MS-based metabolite profiling workflow for any horticultural crop. Sample types can vary depending on the purpose of the study. Often, fruit itself or flower, or any other plant materials, including leaf and root, can be analyzed using GC-MS to capture the data required for an experiment. Regardless of the type of metabolites that need to be analyzed, the first sample preparation step is inhibiting or slowing the active metabolism in a sample. This is usually done using liquid nitrogen or extremely low temperatures to reduce the enzyme activity. This step is known as quenching. After quenching, samples are usually kept at −80 °C until further sample preparation and analysis of samples. The next step in the workflow depends on the types of metabolites, as shown in Fig. 1, including direct analysis of volatile compounds and derivatization methods for the analysis of nonvolatile compounds. In the section below, we provide step-by-step descriptions for selected volatile and nonvolatile metabolite profiling methodologies used in horticultural crops.

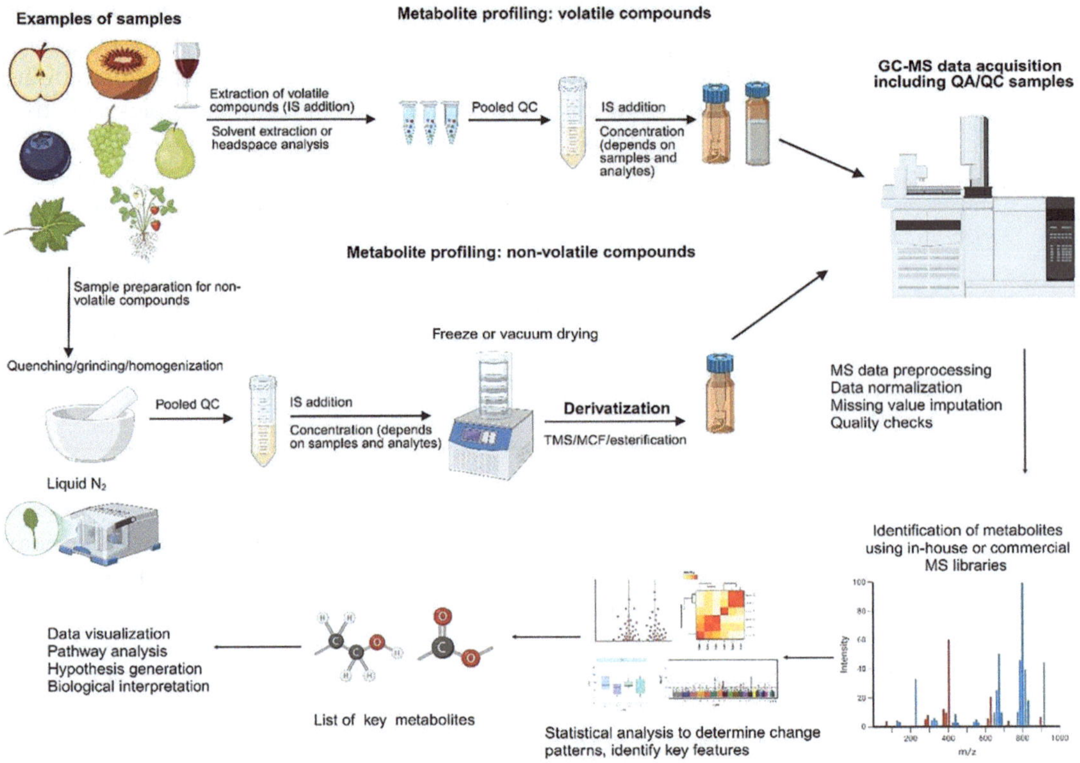

Fig. 1 General overview of GC-MS-based metabolite profiling workflow used for horticulture crops. Here, the steps involved in the analysis of volatile and nonvolatile metabolites are shown. IS = internal standard, QC = quality control, QA = quality assurance, TMS = trimethyl silyl, MCF= methyl chloroformate

3.1 Volatile Profiling

Volatile compounds in a horticultural crop matrix (e.g., fruit pulp, skin, plant leaves, and roots) can be analyzed either directly from the headspace or using a solvent extraction protocol to extract the volatiles of interest.

3.1.1 Solvent-Free Volatile Profiling Method

Characterizing volatile compounds is one of the most common applications of GC-MS in horticultural research. One of the most popular methodologies to analyze volatiles from headspace is using solid phase micro-extraction (SPME), which is a solvent-free method. Here, we provide description of a HS-SPME method used to analyze apple samples to explain the molecular basis for estragole production and its contribution to ripe apple flavor and aroma [64].

1. In liquid nitrogen, 1 to 3 g of the sample was ground to a fine powder.
2. An amount of 30% (w/w) NaCl was added to each sample followed by the addition of internal standard cyclohexanone (2.025 μg/sample) or cetane standard (required for semi-quantification of volatiles).

3. The SPME fiber (1 cm, 50/30 μm, DVB/CAR/PDMS) was exposed to vial headspace to absorb volatiles under the control of a multipurpose sampler injection system (using a PAL RTC robotic sample handler system with an in-line agitator: Gerstel, http://www.gerstel.com/).
4. The vials containing the samples were agitated for 10 minutes at a temperature of 45 °C with agitation at 300 rpm.
5. GC-MS analysis was performed on an HP6890 GC (Agilent) coupled to a time-of-flight mass spectrometer (Leco Pegasus III). SPME fibers were injected into the GC injection port at 240 °C. Injections were splitless for 1 min with a column flow of 1.3 ml/min, followed by 1.1 ml/min in split mode. The GC oven ramp was 45 °C for 1 min, 6 °C/min to 100 °C, 10 °C/min to 150 °C, and 15 °C/min to 240 °C; then held at 240 °C for 5 min.
6. The qualitative and semiquantitative characterization of volatiles was achieved by examining individual m/z 93 ions, the National Institute of Standards and Technology (NIST) database, and comparing retention indices of authentic standards with literature values. Each chemical was semi-quantified using the peak area of its distinctive molecular ion versus total ion area (determined by the Leco software). Concentrations were obtained by comparison with an internal standard of cyclohexanone added to each sample.

Although this study focused on determining the changes in a targeted class of volatiles (estragole), the data generated here provide insights on over hundreds of volatile features, thus making it suitable for volatile profiling of apple samples.

3.1.2 Solvent Extraction Method

Extraction of volatile compounds using suitable solvent/s is another commonly used approach that allows the determination of wide ranges of volatiles present in the horticultural crops. Below is the description of a solvent extraction-based method that has been used to analyze volatile compounds in balsamic vinegars to investigate the differences between vinegar samples based on their origin and potentially quality [65].

1. In a 10 mL glass test tube, 2 mL of balsamic vinegar was added, followed by 1 mL of ether. Then the tube was closed tightly with a screw cap.
2. An amount of 1 mM internal standard 12-bromo-1-dodecanol was added to each sample (5 μL/sample).
3. The vials containing the samples and the internal standards were vortexed for 2 min.

4. Samples were sonicated at 60 Hz for 30 seconds using a sonicator (Soniclean 120 T, Thebarton S.A., Australia) for better extraction of the volatile compounds.
5. The vials were centrifuged at 2000 rpm for 2 min to separate the organic and aqueous layers.
6. Approximately 800 μL of the ether layer was transferred to a silanized glass tube, and the samples were concentrated with N_2 gas to a final volume of approximately 80 μL.
7. Concentrated samples were analyzed using an Agilent GC 7890 coupled to an MSD 5975 (Agilent Technologies, USA) and a quadrupole mass spectrometry selective detector (EI). Samples (1 μL) were injected into the GC injection port at 250 °C. Injections were splitless for 1 minute, followed by a column flow rate of 20 mL/min in split mode. Solvent delay time is 3.8 minutes. The GC oven program was 25 °C for 2 min, 15 °C/min to 280 °C, then held 2 min at 280 °C.
8. Volatile compounds were identified using an in-house MS library that contained mass spectra and retention time index of 300 compounds. In case of unknown compounds not identified using the in-house MS library, the NIST library was used. Data were processed and analyzed using an in-house R package.

The total GC-MS runtime for this method was around 20 min; therefore only highly volatile compounds were the metabolites of interest. Using this method, authors detected and identified over 50 volatile compounds present in different commercial balsamic vinegars.

3.2 Derivatization Methods

Derivatization techniques in GC-MS analysis are essential for the analysis of metabolites with low volatility, poor thermal stability, or low detectability. Acylation, alkylation, silylation, and post-column derivatization techniques are often employed in horticultural research to improve the detectability, stability, and separation efficiency of compounds. During the acylation process, reactive hydrogens in the hydroxyl (-OH), sulfhydryl (-SH), and amine (-NH) groups of the compound structure are transformed into esters, thioesters, and amides to achieve desired volatility and analytical stability [66]. Alkylation derivatization techniques involve the use of alkyl or aryl groups to replace the active hydrogen in the metabolite structure, thereby improving the interaction between the compound and the stationary phase [67, 68]. In addition, the introduction of alkyl groups improves the thermal stability of compounds, such as carboxylic acids, alcohols, phenols, thiols, amides, and sulfonamides, thereby avoiding degradation of the substance to be measured in the GC-MS analysis [69]. Silylation derivatization improves the analytical efficiency of non-polar chromatographic systems by increasing the lipophilicity of the compound by introducing a silyl group (-SiR_3) [70]. Depending on the analytical objective, commonly used silylation reagents include N-methyl-N (trimethylsilyl)trifluoroamide (MSTFA), tert-butyldimethylsilyl (TBDMS), and N,O-bis(trimethylsilyl)trifluoroacetamide (BSTFA) [71]. Unlike other derivatization techniques, post-column derivatization of metabolites occurs after they have been separated by a GC system [72]. Derivative reagents, such as methanol, diazomethane, trifluoroacetyl, and trimethylsilyl (TMS), can be used to improve the ionization efficiency and responsiveness of the isolated compounds or to impart a specific fluorescence or absorbance to the metabolite upon reaction [73]. By modifying the chemical structure to improve separation and detection performance, derivatization remains an indispensable tool in metabolite characterization. As research technology continues to push the boundaries of analytical capability, GC-MS derivatization techniques are becoming faster, more selective, and automated [74].

3.2.1 TMS Derivatization

TMS derivatization methods are among the most widely used protocols in GC-MS based metabolite profiling. For example, TMS derivatization was used to analyze Sauvignon blanc grape juices to determine compositional differences among juices produced in different seasons. This study was the first metabolomics application in grape and wine, where TMS derivatization captured sugars, sugar alcohols, sugar amines, and their derivatives in grape juice samples [74].

1. An amount of 20 μL defrosted grape juice was mixed with 60 μL of methanol, followed by the addition of 20 μL of d-ribitol (internal standard).
2. Using a rotary vacuum dryer (Thermo Fisher, Holbrook, NY, USA), the samples were completely dried to eliminate water.
3. An amount of 80 μL methoxyamine hydrochloride solution in pyridine (2 mg/100 mL) was added to the completely dried grape juice samples and mixed vigorously for 1 min by vortexing to allow resuspension.
4. The resuspended juice mixtures were transferred to a GC vial and incubated for 90 min at 30 °C.
5. An amount of 80 μL of MSTFA was added to the samples and further incubated for 30 min at 37 °C prior to GC–MS analysis.
6. GC–MS analysis of TMS-derivatized samples was performed using an Agilent GC 7890 coupled to an MSD 5975 (Agilent Technologies, USA) with a quadrupole mass selective detector (EI) operated at 70 eV. A Zebron ZB-1701 (Phenomenex), 30 m × 250 μm (internal diameter) × 0.15 μm (film thickness), with a 5 m guard column, was also used for the analysis. The MS was operated in scan mode, where scanning started after

5 min and proceeded through mass range 40–650 a.m.u at 1.47 scans/s.

7. GC-MS data were processed using an in-house R library [62, 75].
8. Metabolites were identified using in-house MS libraries containing information about retention time and mass fragmentation of over 300 metabolite standards.

As part of a metabolomics study on Sauvignon blanc grape juices and wines, this method generated valuable data related to major and minor sugars, sugar alcohols, sugar amines, and their derivatives, complementing data on other primary metabolites, as shown below.

3.2.2 Methyl Chloroformate (MCF) Derivatization

MCF derivatization protocol has already been used for the analysis of grapes, wines, beers, balsamic vinegars, and pomaces from different types of fruit to determine metabolites belonging to amino and non-amino organic acids, and some primary amines and alcohols [62, 75]. An example of MCF derivatization protocol that has been published for metabolite profiling of grape juice and wine samples is provided below [74].

1. A deuterated internal standard (L-alanine-2,3,3,3-d_4, 10 mM, 20 μL) was added to 500 μL of defrosted grape juice/wine samples in a centrifuge tube, mixed well, and frozen.
2. Centrifuged tubes containing frozen samples were placed in a freeze dryer until the mixture was thoroughly dried.
3. The dried samples were reconstituted with methanol (167 μL), pyridine (34 μL), and sodium hydroxide solution (1%, 200 μL), then mixed thoroughly prior to transferring to 5-mL silanized reaction tubes.
4. An amount of 20 μL of MCF was added to the tube and vortexed for 30 seconds.
5. Another 20 μL of MCF was added and again vortexed for 30 seconds.
6. Chloroform (400 μL) was added to the mixture and vortexed for 10 seconds.
7. Sodium bicarbonate solution (50 mM, 400 μL) was added and mixed for another 10 seconds.
8. The mixture was stratified; the upper phase was discarded, while the lower phase was retained.
9. A small amount of anhydrous sodium sulfate was added to dry the chloroform phase, which was then transferred to a salinized GC vial ready for GC-MS analysis.

10. GC–MS analysis was performed using a gas chromatograph GC7890 (Agilent Technologies, USA) coupled to a quadrupole mass spectrometer MSD 5975 (Agilent Technologies, USA) with a ZB-1701 column (30 m × 250 μm id × 0.15 μm film thickness, Phenomenex). The MS operated in scan mode, starting after 5 min and scanning the mass range 40–650 a.m.u at 1.47 scans/s.
11. GC-MS data were processed using an in-house R library [62, 75].
12. Metabolites were identified using in-house MS libraries containing retention time and mass fragmentation data for over 300 metabolite standards.

This method was highly useful as it allowed the determination of metabolites related to central carbon metabolism, thus generating new insights into the composition of Sauvignon blanc juices and wines. The presence of long-overlooked compound classes like fatty acids and other related compounds allowed the generation of hypotheses on the evolution of major aroma compounds and their relationship with many primary metabolites that were not known before.

3.2.3 Fatty Acid Esterification

Fatty acid profiling is another widely used metabolite profiling approach that has been applied to different horticultural crops to understand the composition and evolution of this metabolite class. We provide an example of fatty acid esterification method used for analyzing fatty acid composition of Sauvignon blanc and Pinot noir grapes to understand the impact of ripening and different extraction protocols used during wine production [76]. In this study, different types of samples, including grape juice and alcoholic extracts, seeds, flesh, and skin, were analyzed. Therefore, optimization of sample preparation was first carried out to determine the exact amount of each sample type for the determination of the maximum number of fatty acids within those samples.

1. Based on optimization, 10 mg of sample was used for seeds, skin, and pulp analysis, while 2 mL was used for juice and alcoholic extracts.
2. Grape seeds, skins, and pulp were freeze-dried and ground into powder for lipid extraction.
3. Lipids were extracted by adding 250 μL distilled water to 2 mL juice/alcoholic extracts or 10 mg powder of seeds, skins, and pulp. Chloroform (125 μL) with internal standard (C23:0, 0.48 mg in 50 mL chloroform) and 250 μL methanol were added and mixed thoroughly.

4. Samples were centrifuged at 2500 rpm for 5 min and transferred to new glass vials. Another 250 μL of distilled water and 250 μL of chloroform were added, mixed, and centrifuged again.
5. The top layer was discarded; the bottom chloroform layer containing lipids was transferred to a GC vial and concentrated using a SpeedVac prior to transesterification.
6. The lipid extract (1 mL) was transferred to screw-cap glass tubes, and 2 mL of internal standard (C19:0) solution in methanol:toluene 4:1 (v/v) was added. A 1:10 dilution was applied to seed samples due to their higher lipid content.
7. A magnetic stir bar was placed inside, and 200 μL acetyl chloride was added over 1 min.
8. Tubes were tightly closed and incubated at 100 °C for 1 h in a heating/stirring dry block.
9. Tubes were cooled, dried, and weighed to check for leakage.
10. A volume of 5 mL of 6% K_2CO_3 solution was added slowly, vortexed, and centrifuged at 2500 rpm for 5 min to stop and neutralize the reaction.
11. Approximately 200 μL of the upper toluene phase was transferred to a vial for GC-MS analysis.
12. GC-MS analysis was conducted using an Agilent GC 7890 coupled to an MSD 5975 (Agilent Technologies, Santa Clara, CA, USA) with a quadrupole mass selective detector operated at 70 eV with helium carrier gas. The column was a fused silica Rtx-2330, 100 m long, 0.25 mm internal diameter, 0.2 μm highly polar stationary phase (Shimadzu, Kyoto, Japan).
13. Samples (1 μL) were injected using a CTC PAL autosampler into a glass 4-mm ID straight inlet liner packed with deactivated glass wool. The inlet temperature was 250 °C in splitless mode; column flow was 1 mL/min with a head pressure of 9 psi. Purge flow was 50 mL/min starting 1 min after injection. The oven temperature started isothermally at 45 °C for 2 min, then ramped to 215 °C at 10 °C/min, held 35 min, ramped to 250 °C at 40 °C/min, and held 10 min.
14. The transfer line was maintained at 250 °C, the MSD source at 230 °C, and the quadrupole at 150 °C. The detector was turned on 14.5 min into the run in positive-ion EI mode at 70 eV. Data were acquired at 1463 amu/s in scan mode from 41 to 420 amu, with a detection threshold of 100 ion counts.
15. Fatty acid methyl ester (FAME) peaks were identified by comparison with 52 authentic FAME standards and in-house MS libraries. Quantification used a five-point calibration curve.

This study provided novel knowledge on the changes of saturated and unsaturated fatty acids that change depending on the ripening stages of the grape berries. The authors described a variety of specific compositional differences. Moreover, certain fatty acids and total lipids correlated highly with other traditionally used ripeness parameters, indicating the potential of using fatty acids or lipids as ripening-related biomarkers. Moreover, the extraction of fatty acids in ethanolic solutions relevant to industrial winemaking practices provided knowledge on how aroma compounds can be manipulated by changing the availability of fatty acids during different stages of winemaking.

Considerations that need to be accounted for while preparing and analyzing the samples. In addition, the advancements and drawbacks related to data processing and analysis will also be covered.

3.3 Data Cleaning and Analysis

While data processing for a few targeted metabolite analyses by GC-MS is quite straightforward and relatively easy, data generated from metabolite profiling methods requires a significant amount of time and care during the processing step. As the primary aim of metabolite profiling is to capture information on as many metabolites as possible within a given sample, the acquired data may contain hundreds to thousands of features, depending on the metabolite extraction protocol and instrument sensitivity [1, 43]. There is always the possibility of detecting compounds or contaminants that are not actually present in the samples. Although most GC-MS vendors provide proprietary software for data processing, many other software tools and packages have been developed to handle the large datasets produced by GC-MS analysis. Table 2 summarizes recent advancements in data processing and analytical techniques applied to GC-MS-based metabolite profiling of horticultural crops.

Having a solid understanding of the sample matrix and its impact on results is crucial to generating high-quality data (*see* Subheading 4.1). The use of internal standards—ideally multiple deuterated standards representing different metabolite classes—is highly recommended. These internal standards can be used both to normalize the data and to monitor changes that may occur during sample preparation and analysis. Running instrument and solvent blanks alongside each day's samples is also essential to identify and remove contaminant peaks from the final dataset. Once data are processed carefully, checking the reproducibility among replicates should be conducted. This is typically evaluated using the relative standard deviation (RSD), with values below 20% generally considered acceptable for metabolite profiling [77]. Data from pooled quality control samples, analyzed alongside experimental samples, are increasingly used to assess overall data quality. For example, when pooled QC samples cluster together in an unsupervised

Table 2
Application of data analytical and interpretation tools used in studying different horticultural crops

Samples	Tools	Providers	Licensing	Statistical analyses	Data sizes	Main comparisons among	References
Cabbage	SPSS version 20.0	IBM, New York, USA	Commercial	HCA, PCA, RFC	75 VOCs	10 different cabbage varieties	[79]
Cauliflower, turnip, broccoli, watercress, radish, and cabbage	SIMCA-P version 13.0 software package	Umetrics, Umeå, Sweden	Commercial	HCA, PCA, OPLS-DA	55 primary metabolites	6 edible and nonedible cruciferous vegetables	[80]
Anise, basil, caraway, chamomile, dry coriander, dry mint, fennel, hibiscus, grapes, green beans, green onion, honey, oranges, pomegranate, potatoes, and strawberry	MATLAB program ver. 11.0.	The MathWorks, Natick, MA	Commercial	PCA, ANOVA	200 studied pesticide residues	16 food matrices	[81]
Apple	MATLAB R2015b	The MathWorks, Natick, MA	Commercial	PLS-DA	118 VOCs	42 apple varieties	[82]
Bayberry	Unscrambler v.9.7	CAMO AS, Trondheim, Norway	Commercial	PCA	55 VOCs, 36 aroma-active compounds	3 cultivars	[83]
	SPSS version 17.0	IBM, New York, USA	Commercial	ANOVA			

Date palm	R 2.9.2 software	The R Foundation for Statistical Computing, Vienna, Austria	Open source	PCA, HCA	108 VOCs	13 cultivars	[84]
	SIMCA-P version 13.0 software package	Umetrics, Umeå, Sweden	Commercial	OPLS			
Eggplant	FactoMineR package	Agrocampus Ouest, Rennes, France	Open source	PCA, HCA, heatmap	31 hydrocarbons	36 plant accessions	[85]
Celery	SIMCA software 14	Umetrics, Umeå, Sweden	Commercial	PCA, OPLS–DA	41 VOCs	9 cultivars	[86]
Melon	R language and environment software version 2.14.1 (2014-09-02)	R Foundation for Statistical Computing, Vienna, Austria	Open source	PCA, PLS-DA	137 VOCs	4 cultivars	[87]
Guava	SIMCA-P version 14.1 software package	Umetrics, Umeå, Sweden	Commercial	PCA, HCA, OPLS-DA	42 VOCs	5 cultivars	[88]
Mango	MATLAB R2011A software	The MathWorks, Natick, MA	Commercial	PCA, PLS-DA	108 VOCs	2 physiological statuses	[89]
	R 3.5.2	R Foundation for Statistical Computing, Vienna, Austria	Open source	ANOVA, Levene test, Welch-ANOVA, Pearson's			

(continued)

Table 2 (continued)

Samples	Tools	Providers	Licensing	Statistical analyses	Data sizes	Main comparisons among	References
				correlation coefficient			
Tree tomato	SPSS statistics 25	IBM, New York, USA	Commercial	ANOVA, PCA, PLS-DA, dendrogram, heatmap	46 metabolites	5 ripening stages	[90]
Cocoa	MetaboAnalyst 6.0	Xia lab, McGill University, Canada	Open source	ANOVA, PCA, HCA, dendrogram	63 intrinsic quality markers	27 samples	[91]
Peach	SIMCA version 14.1 software	Umetrics, Umeå, Sweden	Commercial	PCA, OPLS-DA	102 VOCs	142 samples	[92]
Strawberry	Origin2022 pro	OriginLab corporation, Northampton, USA	Commercial	Radar map	92 VOCs	3 ripening stages	[33]
	R 4.3.1 software	R Foundation for Statistical Computing, Vienna, Austria	Open source	Venn diagrams, box plots, heatmaps, volcano plots, and mantel test			

statistical plot, it provides assurance that the data are robust enough to proceed with further statistical analysis. Sample randomization and biological replication should also be incorporated to reduce potential biases that could affect final interpretations [78].

3.4 Metabolite Identification and Quantitation

One of the major bottlenecks of any metabolite profiling and metabolomics studies is the metabolite annotation. However, identification of metabolites by GC-MS is comparatively easier than LC-MS-based metabolite profiling. It is particularly true for volatile compounds as there is usually no derivatization step, thus no or limited chemical changes [43, 70]. Mass spectra of the detected peaks in the samples can be easily identified by comparing mass spectra from known metabolites in commercial MS libraries, including NIST, Human Metabolome Database (HMDB), and MassBank of North America (MoNA), etc. However, many research groups have their own in-house MS library that contains mass spectral information of reference standards and their retention time (RT) index. It is important to consider the elution time of a known compound in the column to avoid any misidentification, which can significantly impact final data interpretation.

For the derivatized metabolites, it is better to rely on in-house MS libraries whenever possible as derivatization changes the chemical structure of the compound and may create multiple reaction products. Commercial databases may not have the required information needed for accurate identification. However, not all the metabolites and their derivatives present in the samples can be identified by the in-house libraries, and the presence of unknown compounds is a major cause of concern, which can be a limiting factor for the overall data interpretation. Recently, many strategies have been proposed to improve metabolite identification by any MS technologies [78, 93, 94]. GC-MS users are also considering minimum reporting standards to avoid false positive or negative identifications.

Usually, metabolite profiling data are semiquantitative in nature where peak height or area of a metabolite is normalized by that of internal standard/s. Many researchers make use of calibration curves of available reference standards if absolute quantification is needed for a particular study. This can be time consuming and doable for targeted analysis of small number of compounds. Recently GC-MS based method has been developed that allowed calibration curve free quantification of primary metabolites [95].

4 Notes

GC-MS data of volatile and nonvolatile compounds contribute to different areas of horticultural sciences including the development of new crop varieties with optimum flavor, identification of unique traits that consumers like, and the reduction of off-flavor and related metabolites. Moreover, GC-MS data can also be used to understand the impact of climate change on the overall quality of fruit/crops and how different processing (microbial fermentation or other industrial processing technologies) led to the development of different products. Considering the multitude of applications shown in this chapter, it is safe to say that GC-MS metabolite profiling methodologies are widely used for the analysis of different horticultural fresh and processed products. However, there are still scopes for improvement, particularly in the data processing and metabolite identification steps. GC-MS applications in horticultural science have both fundamental and applied implications. With time, advancements in this field will bring new innovations, allowing the generation of more data and development of a data analytics platform (e.g. artificial intelligence tools) that ultimately will enable horticulturists to make smart decisions about the crops and their demand by the consumers.

4.1 Matrix Effect

Although GC-MS-based analytical tools are currently among the most powerful analytical tools for horticultural and food research, a major limitation is their susceptibility to matrix effects. Interactions between analytes and components of complex food matrices, such as fruits and vegetables, can cause signal suppression or enhancement, resulting in inaccurate quantification of target analytes. Prior to sample preparation and analysis by GC-MS, it is important to understand the overall matrix of a given sample. For example, horticultural crops like fruits usually contain high concentrations of sugars and organic acids. Depending on the type of metabolites being analyzed, a clean-up step is often required to remove unwanted components that could interfere with the signals of targeted metabolites. During metabolite profiling, fruit samples (e.g., juice) are commonly diluted with appropriate solvents to reduce interference from sugars and acids. Additionally, internal and external standards are typically added to account for any matrix interference [96]. When injecting samples into the GC column, selecting between split and splitless injection modes is crucial for minimizing matrix effects. Furthermore, pooled quality control samples and blank samples prepared with the same solvents and reagents should be analyzed. During data processing, interfering peaks introduced during extraction and derivatization can be identified and subtracted to mitigate matrix impact in the final datasets [1, 43].

References

1. Beale DJ, Pinu FR, Kouremenos KA, Poojary MM, Narayana VK, Boughton BA, Kanojia K, Dayalan S, Jones OA, Dias DA (2018) Review of recent developments in GC–MS approaches to metabolomics-based research. Metabolomics 14:1–31
2. Pinu FR (2018) Grape and wine metabolomics to develop new insights using untargeted and targeted approaches. Fermentation 4:92
3. Villas-Bôas SG, Rasmussen S, Lane GA (2005) Metabolomics or metabolite profiles? Trends Biotechnol 23:385–386
4. Sarabia LD, Hill CB, Boughton BA, Roessner U (2018) Advances of metabolite profiling of plants in challenging environments. In: Annual plant reviews online, pp 629–674
5. Prodhomme D, Valls Fonayet J, Hévin C, Franc C, Hilbert G, de Revel G, Richard T, Ollat N, Cookson SJ (2019) Metabolite profiling during graft union formation reveals the reprogramming of primary metabolism and the induction of stilbene synthesis at the graft interface in grapevine. BMC Plant Biol 19:1–12
6. Vermathen M, Marzorati M, Diserens G, Baumgartner D, Good C, Gasser F, Vermathen P (2017) Metabolic profiling of apples from different production systems before and after controlled atmosphere (CA) storage studied by 1H high resolution-magic angle spinning (HR-MAS) NMR. Food Chem 233:391–400
7. Leng F, Duan S, Song S, Zhao L, Xu W, Zhang C, Ma C, Wang L, Wang S (2021) Comparative metabolic profiling of grape pulp during the growth process reveals systematic influences under root restriction. Metabolites 11:377
8. Lim S, Lee JG, Lee EJ (2017) Comparison of fruit quality and GC–MS-based metabolite profiling of kiwifruit 'Jecy green': natural and exogenous ethylene-induced ripening. Food Chem 234:81–92
9. Xi B-N, Zhang J-J, Xu X, Li C, Shu Y, Zhang Y, Shi X, Shen Y (2024) Characterization and metabolism pathway of volatile compounds in walnut oil obtained from various ripening stages via HS-GC-IMS and HS-SPME-GC–MS. Food Chem 435:137547
10. Hanifah A, Maharijaya A, Putri SP, Laviña WA, Sobir S (2018) Untargeted metabolomics analysis of eggplant (Solanum Melongena L.) fruit and its correlation to fruit morphologies. Metabolites 8:49
11. Farag MA, Maamoun AA, Ehrlich A, Fahmy S, Wesjohann LA (2017) Assessment of sensory metabolites distribution in 3 cactus Opuntia ficus-indica fruit cultivars using UV fingerprinting and GC/MS profiling techniques. LWT 80:145–154
12. Gomez-Gonzalez S, Ruiz-Jimenez J, Priego-Capote F, Luque de Castro MaD (2010) Qualitative and quantitative sugar profiling in olive fruits, leaves, and stems by gas chromatography– tandem mass spectrometry (GC-MS/MS) after ultrasound-assisted leaching. J Agric Food Chem 58:12292–12299
13. Uri C, Juhász Z, Polgár Z, Bánfalvi Z (2014) A GC–MS-based metabolomics study on the tubers of commercial potato cultivars upon storage. Food Chem 159:287–292
14. Niu E, Hu W, Ding J, Wang W, Romero A, Shen G, Zhu S (2022) GC-MS/LC-MS and transcriptome analyses revealed the metabolisms of fatty acid and flavonoid in olive fruits (Olea europaea L.). Sci Hortic 299:111017
15. Adjepong M, Valentini K, Pickens CA, Li W, Appaw W, Fenton J (2017) Quantification of fatty acid and mineral levels of selected seeds, nuts, and oils in Ghana. J Food Compos Anal 59:43–49
16. Abed SS, Kiranmayi P, Imran K, Lateef SS (2023) Gas chromatography-mass spectrometry (GC-MS) metabolite profiling of Citrus Limon (L.) Osbeck juice extract evaluated for its antimicrobial activity against Streptococcus mutans. Cureus 15:e33585
17. Vrhovsek U, Lotti C, Masuero D, Carlin S, Weingart G, Mattivi F (2014) Quantitative metabolic profiling of grape, apple and raspberry volatile compounds (VOCs) using a GC/MS/MS method. J Chromatogr B 966: 132–139
18. Nam KH, Kim DY, Kim HJ, Pack I-S, Kim HJ, Chung YS, Kim SY, Kim C-G (2019) Global metabolite profiling based on GC–MS and LC–MS/MS analyses in ABF3-overexpressing soybean with enhanced drought tolerance. Appl Biol Chem 62
19. Razzaq A, Sadia B, Raza A, Hameed MK, Saleem F (2019) Metabolomics: a way forward for crop improvement. Metabolites 9:303
20. Swamy MK, Arumugam G, Kaur R, Ghasemzadeh A, Yusoff MM, Sinniah UR (2017) GC-MS based metabolite profiling, antioxidant and antimicrobial properties of different solvent extracts of Malaysian Plectranthus amboinicus leaves. Evid Based Complement Alternat Med 2017:1517683
21. Zeki ÖC, Eylem CC, Reçber T, Kır S, Nemutlu E (2020) Integration of GC–MS and LC–MS

for untargeted metabolomics profiling. J Pharm Biomed Anal 190:113509
22. Gou M, Bi J, Chen Q, Wu X, Fauconnier M-L, Qiao Y (2023) Advances and perspectives in fruits and vegetables flavor based on molecular sensory science. Food Rev Intl 39:3066–3079
23. Hagos M, Yaya EE, Chandravanshi BS, Redi-Abshiro M (2022) Analysis of volatile compounds in flesh, peel and seed parts of pumpkin (Cucurbita maxima) cultivated in Ethiopia using gas chromatography-mass spectrometry (GC-MS). Int J Food Prop 25:1498–1512
24. Choi B-S, Lee D-U, Kim W-S, Park C-W, Choe W-J, Moon M-J (2023) Simultaneous screening of 322 residual pesticides in fruits and vegetables using GC-MS/MS and deterministic health risk assessments. Foods 12:3001
25. Kasal-Slavik T, Eschweiler J, Kleist E, Mumm R, Goldbach HE, Schouten A, Wildt J (2017) Early biotic stress detection in tomato (Solanum lycopersicum) by BVOC emissions. Phytochemistry 144:180–188
26. Oliete B, Lubbers S, Fournier C, Jeandroz S, Saurel R (2022) Effect of biotic stress on the presence of secondary metabolites in field pea grains. J Sci Food Agric 102:4942–4948
27. Shen X, Wang Y, Ran L, Liu R, Sun X, Hu L, Xiao Y, Chen F (2022) Flavor deterioration of liquid endosperm in postharvest tender coconut revealed by LC-MS-based metabolomics, GC-IMS and E-tongue. Postharvest Biol Technol 187:111866
28. Terletskaya NV, Shadenova EA, Litvinenko YA, Ashimuly K, Erbay M, Mamirova A, Nazarova I, Meduntseva ND, Kudrina NO, Korbozova NK, Djangalina ED (2024) Influence of cold stress on physiological and phytochemical characteristics and secondary metabolite accumulation in microclones of Juglans Regia L. Int J Mol Sci 25:4991
29. Hou J, Liang L, Wang Y (2020) Volatile composition changes in navel orange at different growth stages by HS-SPME–GC–MS. Food Res Int 136:109333
30. Li H, Brouwer B, Oud N, Verdonk JC, Tikunov Y, Woltering E, Schouten R, da Silva FP (2021) Sensory, GC-MS and PTR-ToF-MS profiling of strawberries varying in maturity at harvest with subsequent cold storage. Postharvest Biol Technol 182:111719
31. Wu S, Yin J, Li X, Xie J, Ding H, Han L, Bie S, Li F, Zhu B, Kang L (2023) An exploration of dynamic changes in the mulberry growth process based on UPLC-Q-Orbitrap-MS, HS-SPME-GC-MS, and HS-GC-IMS. Foods 12:3335
32. Lee K-M, Oh TI, Kim S-H, Kim H-Y, Chung HS, Min DS, Auh J-H, Lee HJ, Lee J, Choi H-W (2016) Comprehensive metabolic profiles of mulberry fruit (Morus Alba Linnaeus) according to maturation stage. Food Sci Biotechnol 25:1035–1041
33. Liu H, Wu S, Liu M, Sun Y, Li J (2024) Exploring flavour profile changes of strawberry fruits under different organic fertilizer strategies by combining hs-spme-gc-ms and metabolomics. J Food Compos Anal 135:106653
34. Bekele EA, Annaratone CE, Hertog ML, Nicolai BM, Geeraerd AH (2014) Multi-response optimization of the extraction and derivatization protocol of selected polar metabolites from apple fruit tissue for GC–MS analysis. Anal Chim Acta 824:42–56
35. Zuo Y, Wang C, Zhan J (2002) Separation, characterization, and quantitation of benzoic and phenolic antioxidants in American cranberry fruit by GC– MS. J Agric Food Chem 50:3789–3794
36. Topolewska A, Czarnowska K, Haliński ŁP, Stepnowski P (2015) Evaluation of four derivatization methods for the analysis of fatty acids from green leafy vegetables by gas chromatography. J Chromatogr B 990:150–157
37. Albero B, Sánchez-Brunete C, Miguel E, Tadeo JL (2017) Application of matrix solid-phase dispersion followed by GC–MS/MS to the analysis of emerging contaminants in vegetables. Food Chem 217:660–667
38. Qin G, Zou K, Tian L, Li Y (2018) Determination of five plant growth regulator containing carboxyl in bean sprouts based on chemical derivatization by GC-MS. Food Anal Methods 11:2628–2635
39. George J, Nguyen TT, Sanewski G, Hardner C, Smyth HE (2024) Stable isotope dilution assay and HS-SPME-GCMS quantification of key aroma volatiles of Australian pineapple (Ananas comosus) cultivars. Food Chem 455:139956
40. Pook C, Diep TT, Yoo MJY (2022) Simultaneous quantification of organic acids in tamarillo (Solanum betaceum) and untargeted chemotyping using methyl chloroformate derivatisation and GC-MS. Molecules 27:1314
41. Khakimov B, Mongi RJ, Sørensen KM, Ndabikunze BK, Chove BE, Engelsen SB (2016) A comprehensive and comparative GC–MS metabolomics study of non-volatiles in Tanzanian grown mango, pineapple, jackfruit, baobab and tamarind fruits. Food Chem 213:691–699
42. Abdelmonsef M, Shawky E, Ghareeb DA, El Newehy NM (2024) Comprehensive metabolomics and chemometrics unravel potential

anti-diabetic metabolites of pumpkin (Cucurbita pepo L.) fruits through UPLC-QqQ-MS and GC–MS analyses. Food Res Int 192: 114771

43. Beale DJ, Kouremenos KA, Palombo EA (2016) Microbial metabolomics: applications in clinical, environmental, and industrial microbiology. In: Microbial metabolomics: applications in clinical, environmental, and industrial microbiology. Springer International Publishing. https://doi.org/10.1007/978-3-319-46326-1
44. Applequist WL, Miller JS (2013) Selection and authentication of botanical materials for the development of analytical methods. Anal Bioanal Chem 405:4419–4428
45. Akili AWR, Hardianto A, Latip J, Ismiyati M, Herlina T (2023) Characterization of botanical parts of Erythrina crista-galli using pyrolysis-gas chromatography/mass spectrometry and multivariate analysis. Indonesian J Chem 23: 899
46. Demırcı B, DİLmaÇ E, Kırcı D, Ilhan G, Demirci F, Kılıç CS, Duman H (2020) Chemical and antimicrobial characterization of essential oils obtained from aerial part, root and fruit of Ferulago longistylis Boiss., an endemic species. Nat Volat Essent Oils 7:18–25
47. Ashraf MA, Iqbal M, Rasheed R, Hussain I, Riaz M, Arif MS (2018) Environmental stress and secondary metabolites in plants: an overview. In: Plant metabolites and regulation under environmental stress, pp 153–167
48. Pant P, Pandey S, Dall'Acqua S (2021) The influence of environmental conditions on secondary metabolites in medicinal plants: a literature review. Chem Biodivers 18:e2100345
49. Bonat Celli G, Ghanem A, Su-Ling Brooks M (2016) Influence of freezing process and frozen storage on the quality of fruits and fruit products. Food Rev Intl 32:280–304
50. Palta JP, Weiss LS (2018) Ice formation and freezing injury: an overview on the survival mechanisms and molecular aspects of injury and cold acclimation in herbaceous plants. In: Advances in plant cold hardiness, pp 143–176
51. Bouzari A, Holstege D, Barrett DM (2015) Vitamin retention in eight fruits and vegetables: a comparison of refrigerated and frozen storage. J Agric Food Chem 63:957–962
52. Dias MG, Camões MFG, Oliveira L (2014) Carotenoid stability in fruits, vegetables and working standards–effect of storage temperature and time. Food Chem 156:37–41
53. Zhan X, Zhu Z, Sun D-W (2019) Effects of pretreatments on quality attributes of long-term deep frozen storage of vegetables: a review. Crit Rev Food Sci Nutr 59:743–757
54. Cao X, Zhang F, Zhao D, Zhu D, Lǐ J (2018) Effects of freezing conditions on quality changes in blueberries. J Sci Food Agric 98: 4673–4679
55. Bartosińska E, Buszewska-Forajta M, Siluk D (2016) GC–MS and LC–MS approaches for determination of tocopherols and tocotrienols in biological and food matrices. J Pharm Biomed Anal 127:156–169
56. Grauwet T, Kebede BT, Delgado RM, Lemmens L, Manzoni F, Vervoort L, Hendrickx M, Stephen Elmore J, Van Loey A (2015) Evaluating the potential of high pressure high temperature and thermal processing on volatile compounds, nutritional and structural properties of orange and yellow carrots. Eur Food Res Technol 240:183–198
57. Kaulmann A, André CM, Schneider Y-J, Hoffmann L, Bohn T (2016) Carotenoid and polyphenol bioaccessibility and cellular uptake from plum and cabbage varieties. Food Chem 197:325–332
58. Liu Y, Zhang Z, Hu L (2022) High efficient freeze-drying technology in food industry. Crit Rev Food Sci Nutr 62:3370–3388
59. Steingass CB, Carle R, Schmarr H-G (2015) Ripening-dependent metabolic changes in the volatiles of pineapple (Ananas comosus (L.) Merr.) fruit: I. Characterization of pineapple aroma compounds by comprehensive two-dimensional gas chromatography-mass spectrometry. Anal Bioanal Chem 407:2591–2608
60. Mihaylova D, Popova A, Vrancheva R, Dincheva I (2022) HS-SPME-GC–MS volatile profile characterization of peach (Prunus persica L. Batsch) varieties grown in the eastern Balkan Peninsula. Plants 11:166
61. Duportet X, Aggio RBM, Carneiro S, Villas-Bôas SG (2012) The biological interpretation of metabolomic data can be misled by the extraction method used. Metabolomics 8: 410–421
62. Tumanov S, Zubenko Y, Greven M, Greenwood DR, Shmanai V, Villas-Boas SG (2015) Comprehensive lipidome profiling of sauvignon blanc grape juice. Food Chem 180: 249–256
63. Kvitvang HF, Andreassen T, Adam T, Villas-Bôas SG, Bruheim P (2011) Highly sensitive

GC/MS/MS method for quantitation of amino and nonamino organic acids. Anal Chem 83:2705–2711

64. Yauk YK, Chagné D, Tomes S, Matich AJ, Wang MY, Chen X, Maddumage R, Hunt MB, Rowan DD, Atkinson RG (2015) The O-methyltransferase gene M do OMT 1 is required for biosynthesis of methylated phenylpropenes in ripe apple fruit. Plant J 82:937–950
65. Pinu FR, De Carvalho-Silva S, Trovatti Uetanabaro AP, Villas-Boas SG (2016) Vinegar metabolomics: an explorative study of commercial balsamic vinegars using gas chromatography-mass spectrometry. Metabolites 6:22
66. Wang W, Wang MY, Zeng Y, Chen X, Wang X, Barrington AM, Tao J, Atkinson RG, Nieuwenhuizen NJ (2023) The terpene synthase (TPS) gene family in kiwifruit shows high functional redundancy and a subset of TPS likely fulfil overlapping functions in fruit flavour, floral bouquet and defence. Mol Horticult 3:9
67. Orata F (2012) Derivatization reactions and reagents for gas chromatography analysis. In: Advanced gas chromatography-progress in agricultural, biomedical and industrial applications, p 91
68. Glicksberg L, Kerrigan S (2020) Derivatization. In: Principles of forensic toxicology, pp 163–175
69. Yang Y, Lin M, Tang J, Ma S, Yu Y (2020) Derivatization gas chromatography negative chemical ionization mass spectrometry for the analysis of trace organic pollutants and their metabolites in human biological samples. Anal Bioanal Chem 412:6679–6690
70. Pinu FR, Villas-Boas SG, Aggio R (2017) Analysis of intracellular metabolites from microorganisms: quenching and extraction protocols. Metabolites 7:53
71. Farajzadeh MA, Nouri N, Khorram P (2014) Derivatization and microextraction methods for determination of organic compounds by gas chromatography. TrAC Trends Anal Chem 55:14–23
72. Lavilla I, Romero V, Costas I, Bendicho C (2014) Greener derivatization in analytical chemistry. TrAC Trends Anal Chem 61:1–10
73. Zarate E, Boyle V, Rupprecht U, Green S, Villas-Boas SG, Baker P, Pinu FR (2017) Fully automated trimethylsilyl (TMS) derivatisation protocol for metabolite profiling by GC-MS. Metabolites 7:1
74. Pinu FR, Edwards PJ, Jouanneau S, Kilmartin PA, Gardner RC, Villas-Boas SG (2014) Sauvignon blanc metabolomics: grape juice metabolites affecting the development of varietal thiols and other aroma compounds in wines. Metabolomics 10:556–573
75. Pinu FR, Tumanov S, Grose C, Raw V, Albright A, Stuart L, Villas-Boas SG, Martin D, Harker R, Greven M (2019) Juice index: an integrated sauvignon blanc grape and wine metabolomics database shows mainly seasonal differences. Metabolomics 15:1–18
76. Sherman E, Yvon M, Grab F, Zarate E, Green S, Bang KW, Pinu FR (2023) Total lipids and fatty acids in major New Zealand grape varieties during ripening, prolonged pomace contacts and ethanolic extractions mimicking fermentation. Fermentation 9:357
77. Pérez-Jiménez M, Sherman E, Pozo-Bayón MÁ, Muñoz-González C, Pinu FR (2023) Application of untargeted volatile profiling to investigate the fate of aroma compounds during wine oral processing. Food Chem 403: 134307
78. Alseekh S, Aharoni A, Brotman Y, Contrepois K, D'Auria J, Ewald J, C. Ewald J, Fraser PD, Giavalisco P, Hall RD (2021) Mass spectrometry-based metabolomics: a guide for annotation, quantification and best reporting practices. Nat Methods 18:747–756
79. Wei S, Xiao X, Wei L, Li L, Li G, Liu F, Xie J, Yu J, Zhong Y (2021) Development and comprehensive HS-SPME/GC–MS analysis optimization, comparison, and evaluation of different cabbage cultivars (Brassica oleracea L. var. capitata L.) volatile components. Food Chem 340:128166
80. Baky MH, Shamma SN, Xiao J, Farag MA (2022) Comparative aroma and nutrients profiling in six edible versus nonedible cruciferous vegetables using MS based metabolomics. Food Chem 383:132374
81. Shendy AH, Eltanany BM, Al-Ghobashy MA, Gadalla SA, Mamdouh W, Lotfy HM (2019) Coupling of GC-MS/MS to principal component analysis for assessment of matrix effect: efficient determination of ultra-low levels of pesticide residues in some functional foods. Food Anal Methods 12:2870–2885
82. Giannetti V, Mariani MB, Mannino P, Marini F (2017) Volatile fraction analysis by HS-SPME/GC-MS and chemometric modeling for traceability of apples cultivated in the Northeast Italy. Food Control 78:215–221
83. Cheng H, Chen J, Chen S, Wu D, Liu D, Ye X (2015) Characterization of aroma-active volatiles in three Chinese bayberry (Myrica rubra) cultivars using GC–MS–olfactometry and an electronic nose combined with principal component analysis. Food Res Int 72:8–15

84. Khalil MN, Fekry MI, Farag MA (2017) Metabolome based volatiles profiling in 13 date palm fruit varieties from Egypt via SPME GC–MS and chemometrics. Food Chem 217: 171–181
85. Haliński ŁP, Samuels J, Stepnowski P (2017) Multivariate analysis as a key tool in chemotaxonomy of brinjal eggplant, African eggplants and wild related species. Phytochemistry 144: 87–97
86. Yan J, Chen J, Huang Z, He L, Wu L, Yu L, Zhu W (2024) Characterisation of the volatile compounds in nine varieties and three breeding selections of celery using GC–IMS and GC–MS. Food Chemistry: X 24:101936
87. Chaparro-Torres LA, Bueso MC, Fernández-Trujillo JP (2016) Aroma volatiles obtained at harvest by HS-SPME/GC-MS and INDEX/MS-E-nose fingerprint discriminate climacteric behaviour in melon fruit. J Sci Food Agric 96: 2352–2365
88. Kamal IM, Zayed A, Eissa TF, Farag MA (2023) Aroma-based discrimination of Egyptian versus Indian guava fruits and in response to probiotics as analyzed via SPME/GC–MS and chemometric tools. Sci Rep 13:18420
89. Oldoni FCA, Florencio C, Bertazzo GB, Grizotto PA, Junior SB, Carneiro RL, Colnago LA, Ferreira MD (2022) Fruit quality parameters and volatile compounds from 'Palmer'-mangoes with internal breakdown. Food Chem 388:132902
90. Ramírez MQ, Valdez EA, Aguirre NC, Duno D, Ocampo GT (2023) Volatilomic profile of the tree tomato (Solanum betaceum Cav.) pulp during ripening and senescence using HS–SPME with GC–MS. LWT 186: 115213
91. Sari ABT, Andriana Y, Abdurrahman MF, Kurnianto D, Sefrienda AR, Setiyoningrum F, Minh TN (2024) Differentiation of Indonesian specialty cocoa (Theobroma cacao) varieties and origins using phenolic contents, antioxidant activities, and spectrometric fingerprints for traceability. Food Biosci 62:105434
92. Giménez-Campillo C, Arroyo-Manzanares N, Pastor-Belda M, Campillo N, Viñas P (2024) Discrimination of the geographical origin of Cacaoes by the monitoring of volatile organic compounds by gas chromatography with mass spectrometry and chemometric tools. J Food Compos Anal 129:106125
93. Zhou Z, Luo M, Zhang H, Yin Y, Cai Y, Zhu Z-J (2022) Metabolite annotation from knowns to unknowns through knowledge-guided multi-layer metabolic networking. Nat Commun 13:6656
94. Fiehn O (2016) Metabolomics by gas chromatography–mass spectrometry: combined targeted and untargeted profiling. Curr Protoc Mol Biol 114:30.34.31–30.34.32
95. Tarakhovskaya E, Marcillo A, Davis C, Milkovska-Stamenova S, Hutschenreuther A, Birkemeyer C (2023) Matrix effects in GC-MS profiling of common metabolites after trimethylsilyl derivatization. Molecules 28
96. Tumanov S, Zubenko Y, Obolonkin V, Greenwood DR, Shmanai V, Villas-Boas SG (2016) Calibration curve-free GC–MS method for quantitation of amino and non-amino organic acids in biological samples. Metabolomics 12: 1–13

Chapter 8

GC-MS-Based Metabolite Profiling Method for Tea

Enik Nurlaili Afifah, Shunsuke Miyauchi, and Eiichiro Fukusaki

Abstract

Tea is one of the most widely consumed beverages in the world, valued not only for its distinctive flavors but also for its potential health benefits. To better understand the quality of tea based on the metabolite profiles, researchers have increasingly employed metabolomic approaches to analyze its chemical composition across various processing methods and tea varieties. Although most studies have focused on targeted metabolomics using liquid chromatography (LC)-based techniques, research utilizing gas chromatography-mass spectrometry (GC-MS) remains limited. Moreover, standardized protocols for applying GC-MS in tea analysis are still lacking. This book chapter aims to address this gap by presenting a GC-MS-based metabolomic workflow, including sample extraction, derivatization, and instrumental parameters for the analysis of nonvolatile metabolites. By offering practical guidance, this chapter supports researchers in using GC-MS as a powerful tool for comprehensive metabolite profiling and expanding chemical coverage in tea research.

Key words Agriculture, GC-MS, Metabolites, Protocol, Profiling, Tea

1 Introduction

Tea, a processed product derived from the leaves *of Camellia sinensis*, is the most widely consumed beverage in the world after water [1]. It is enjoyed in various forms, including green, black, oolong, white, and fermented teas, each offering distinct flavors, aromas, and potential health benefits [2, 3]. Tea is particularly rich in bioactive compounds such as catechins, theanine, and caffeine, which have been associated with antioxidant, anti-inflammatory, and calming effects [4]. In recent years, growing consumer interest in functional foods and beverages has further driven the expansion of the global tea market, with increasing demand for products that combine sensory appeal with health-promoting properties. In parallel with this growing interest, scientific studies on tea quality and its link to health benefits have also intensified [5].

To better understand the quality and functional properties of tea, many studies have focused on its metabolite composition.

Sastia Prama Putri (ed.), *Gas Chromatography-Mass Spectrometry: Based Metabolomics in Food Science*, Methods and Protocols in Food Science, https://doi.org/10.1007/978-1-0716-5108-7_8,

High-quality tea typically exhibits a desirable aroma and taste, which are influenced by a diverse array of metabolites [4, 6, 7] reported that tea quality is shaped by its chemical composition, which can be significantly altered by post-harvest processing techniques as well as the genetic background. Consequently, numerous studies have explored the chemical profiles of different tea types using advanced analytical methods [7, 8]. The flavor and sensory characteristics of tea are generally determined by both volatile and nonvolatile compounds. While volatile metabolites are primarily responsible for aroma, nonvolatile compounds—such as polyphenols, amino acids, and organic acids—play key roles in taste [2, 9].

In addition to their sensory contributions, these chemical constituents are often linked to various health benefits [3]. For instance, tea contains a wide spectrum of bioactive compounds, including flavonoids, amino acids, vitamins, phenolic acids, gallic and chlorogenic acids, lipids, proteins, carbohydrates, β-carotene, and fluoride—many of which have demonstrated pharmacological activities such as antioxidant, antimicrobial, and growth-promoting effects [10]. Accordingly, research on the potential health-promoting functions of tea and its constituents continues to grow [11, 12].

To explore these complex chemical profiles, researchers have employed a range of metabolomics techniques. Modern analytical platforms such as Liquid Chromatography–Mass Spectrometry (LC-MS) and Ultra-High-Performance Liquid Chromatography coupled with Quadrupole Time-of-Flight Mass Spectrometry (UHPLC-qTOF-MS) are frequently used to analyze tea metabolites [13], particularly those associated with health outcomes [11], as well as in relation to the taste quality [14]. However, Gas Chromatography–Mass Spectrometry (GC-MS) has gained prominence as a robust and efficient method for profiling tea metabolites, especially for nontargeted or widely targeted analysis [15, 16]. Typically, researchers first apply GC-MS to identify a broad range of compounds, followed by targeted LC-MS analysis to confirm and quantify specific metabolites [17]. In addition, GC-MS is well-suited for detecting compounds related to the taste and aroma of tea, making it a promising analytical tool in tea metabolomics.

GC-MS offers several advantages, including higher sensitivity to volatile and nonvolatile compounds, simpler data processing, and more accessible metabolite libraries [18, 19]. Nevertheless, despite its growing popularity, standardized protocols for GC-MS-based metabolomics in tea research remain limited. Most existing studies have focused on targeted analyses using LC-MS approaches, while widely targeted GC-MS applications are still emerging [3, 7, 20]. Additionally, the number and types of metabolites detected using GC-MS vary greatly between studies due to differences in instrument conditions—such as the column type, temperature settings, and extraction and derivatization protocols [17, 21].

To improve the reproducibility and accuracy of tea metabolomics research, there is a clear need for detailed and standardized protocols—particularly for GC-MS analysis, including sample preparation, derivatization methods, and instrument parameters. This chapter aims to address that gap by providing a comprehensive guide to GC-MS analysis of tea samples, from sample extraction to optimized instrument settings, thereby supporting researchers in generating high-quality and comparable metabolomics data.

2 Materials

2.1 Sample Type

1. Fresh tea leaves.
2. Black tea.
3. Green tea.
4. Other processed tea samples.

2.2 Chemicals and Reagents

1. Liquid nitrogen.
2. Methanol (MeOH, ≥99.8%).
3. Ultrapure water (H_2O).
4. Chloroform ($CHCl_3$, ≥99%).
5. Ribitol (internal standard).
6. Methoxyamine hydrochloride.
7. Pyridine (≥99.5%).
8. N-Methyl-N-(trimethylsilyl) trifluoroacetamide (MSTFA).
9. Alkane mixture standard (C9–C40).

2.3 Laboratory Supplies

1. Polycarbonate grinding tubes (50 mL).
2. Eppendorf tubes (1.5 mL, 2 mL, and 15 mL).
3. GC-MS vials with caps.
4. Amber glass bottles (100 mL).
5. Micropipette tips.
6. Fused silica capillary column (30 m × 0.25 mm i.d., 0.25 μm film thickness, CP-SIL 8 CB, low bleed; Agilent Technologies).

2.4 Equipment

1. Freeze-dryer.
2. Micropipettes.
3. Spatula.
4. CryoTank.
5. Multi-beads shocker (Yasui Kikai Co., Osaka, Japan).
6. Thermomixer.

7. Vortex mixer.
8. Refrigerated centrifuge.
9. Vacuum concentrator.
10. GC–MS instruments:
    - GCMS-QP2010 Ultra (Shimadzu, Kyoto, Japan)-
    - GCMS-TQ8030 (Shimadzu, Kyoto, Japan)-
    - Autosampler: AOC-20i/s (Shimadzu, Kyoto, Japan)-

2.5 Software

1. GCMSsolution (Shimadzu).
2. Reifycs ABF Converter (https://www.reifycs.com/abfconverter/).
3. MS-DIAL (https://systemsomicslab.github.io/compms/msdial/main.html)
4. GL-Science DB Library—InertCap 5MS-NP, 494 records (https://systemsomicslab.github.io/compms/msdial/main.html#MSP).
5. SIMCA-P (licensed software).
6. NIST 14 Library (National Institute of Standards and Technology).
7. RStudio (https://www.r-project.org).

3 Method

The GC-MS protocol for tea samples was based on previously published methods by Myauchi et al. [15] and Pongsuwan et al. [22] with some modifications. The overall procedures are shown in Fig. 1.

3.1 Sample Preparation

1. If the tea sample is not fully dry, transfer it to a 50 mL polycarbonate tube and seal with a cap.
2. Immerse the tube in liquid nitrogen (*see* **Note 1**).
3. Replace the cap with parafilm, pierce holes, and place it into the freeze-dryer chamber.
4. Dry the sample at −50 °C for 24 h or until fully desiccated (*see* **Note 2**).
5. Transfer ~25 mL of dried tea sample into a grinding tube.
6. Add a metal bead, seal the tube, and dip briefly in liquid nitrogen.
7. Process in a Multi-beads shocker at 2000 rpm for 30 s; repeat 2–3 times until a fine powder is obtained (*see* **Note 3**).

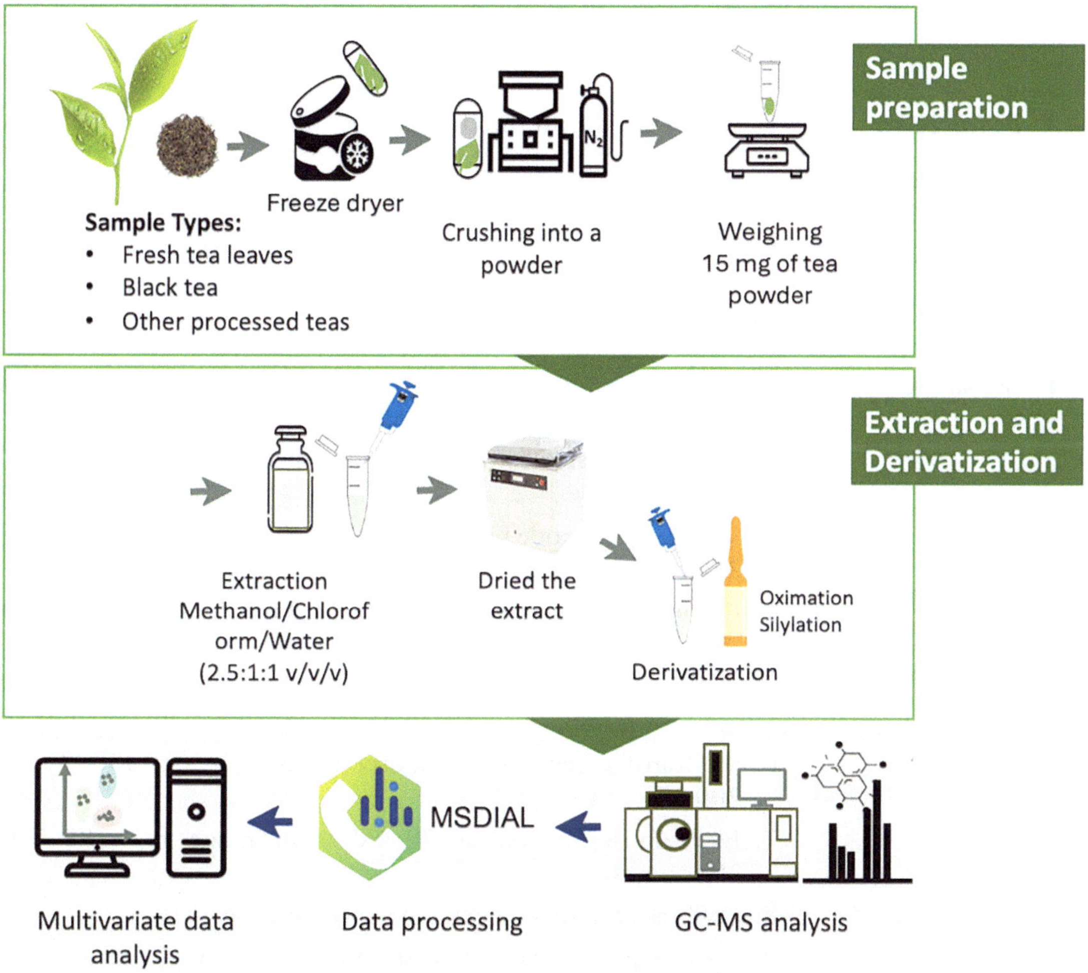

Fig. 1 GC-MS nonvolatile analysis workflow

3.2 Extraction of Polar Metabolites

1. Weigh 15 mg of powdered tea into a 2 mL Eppendorf tube. Prepare three technical replicates ($n = 3$) per sample.
2. Include a blank tube processed identically.
3. Add 1 mL of extraction solvent ($MeOH:H_2O:CHCl_3$, 2.5:1:1, v/v/v) with ribitol (0.02 mg/mL) as internal standard.
4. Vortex for 5 min, incubate at 37 °C, 1200 rpm for 30 min.
5. Centrifuge at 16,000 × g for 3 min at 4 °C.
6. Transfer 900 μL of the supernatant to a new 1.5 mL tube.
7. Add 400 μL of ultrapure water, vortex, and centrifuge again.
8. For QC, pool 100 μL of aqueous layers from each sample (excluding blank).
9. Transfer 100 μL of the upper aqueous phase to a fresh 1.5 mL tube (with pierced cap).

10. Evaporate using a vacuum concentrator for 120 min.
11. Lyophilize overnight.

3.3 Derivatization

1. Add 100 μL of methoxyamine hydrochloride in pyridine (20 mg/mL) to the lyophilized sample.
2. Incubate at 30 °C, 1200 rpm for 90 min.
3. Add 50 μL of MSTFA and incubate at 37 °C for 30 min.
4. Transfer 200 μL of each derivatized sample, QC, and blank to GC-MS vials and seal.

3.4 GC-MS Analysis

1. Analyze using GCMS-QP2010 Ultra or GCMS-TQ8030 equipped with AOC-20i/s autosampler.
2. Prepare a batch file listing vial order, method, injection volume, and tuning file.
3. Set injection mode to split (10:1), with an inlet temperature of 230 °C.
4. Use hydrogen as the carrier gas at 1.12 mL/min; linear velocity 39 cm/s.
5. Oven program: 80 °C (2 min), ramp 15 °C/min to 330 °C, hold for 6 min.
6. Set transfer line and ion source temperatures at 250 °C and 200 °C.
7. Electron ionization at 70 eV; scan range m/z 85–500, 20 scans/s.
8. Calibrate the mass spectrometer before use.
9. Inject 1 μL of alkane standard mixture (C9–C40) before sample analysis for retention index (RI) calculation (*see* **Note 4**).
10. Inject derivatized samples randomly; include a QC injection every 4–5 samples.

3.5 Data Processing and Analysis

1. Convert raw chromatograms to netCDF format using GCMSsolution.
2. Convert netCDF files to ABF format using Reifycs ABF Converter.
3. Import ABF files into MS-DIAL for peak detection and alignment.
4. Set peak detection parameters:
   (a) Minimum peak height: 1000.
   (b) Smoothing: Linear weighted moving average.
   (c) Smoothing level: 5 scans.
   (d) Average peak width: 20 scans.

(e) σ-window value: 0.5.

(f) EI cut-off: 200 amplitude.

5. Identify peaks using alkane RI and GL-Science DB.
6. Set MS-DIAL parameters:
   (a) RI tolerance: 10.
   (b) RT tolerance: 0.5 min.
   (c) m/z tolerance: 0.5 Da.
   (d) EI similarity cut-off: 70%.
   (e) Identification score threshold: 70%.
7. Normalize peak heights against ribitol internal standard.
8. Export data matrix with raw and normalized values.
9. Remove peaks present in blanks (*see* **Note 5**).
10. Exclude peaks with >30% RSD in QC samples.
11. Use the filtered matrix for multivariate analyses (PCA, OPLS-DA, OPLS-R) in SIMCA-P or RStudio.

4 Notes

1. Handling of Liquid Nitrogen: Liquid nitrogen can cause rapid and vigorous boiling when it contacts room-temperature tubes. Always make sure to tightly close the tube immediately after immersion to prevent splashing or partial loss of the sample. Use appropriate personal protective equipment (PPE), including cryo-gloves and a face shield, and handle liquid nitrogen in a well-ventilated area.
2. Ensuring Complete Freeze-Drying: Proper freeze-drying is essential to remove all moisture from the sample. Residual water can interfere with derivatization efficiency and lead to inconsistent GC-MS results. If samples still appear soft or slightly sticky, extend freeze-drying up to 48 hours to ensure full dryness.
3. Efficient Pulverization Using Multi-Beads Shocker: Pulverizing in liquid nitrogen maintains sample integrity and ensures consistent extraction. If the sample is not well powdered after two or three cycles, this may be due to insufficient freezing or the lid not being properly sealed. Re-immerse the tube in liquid nitrogen before repeating the shocker cycles, and confirm the lid is secure to prevent leakage or pressure buildup.
4. Use of Alkane Standard for RI Calibration: Injecting a C9–C40 alkane standard under the same GC conditions is crucial for calculating accurate retention indices (RI). This enhances metabolite identification confidence, particularly when spectral

similarity alone is insufficient. Run the alkane standard at least once per sequence or after long analysis runs.

5. Exclusion of Background Peaks Based on Blank and QC Filtering: Peaks present in blank samples often originate from reagents or column bleed. Features with intensities less than five times those in blanks should be excluded. Additionally, features showing >30% RSD in pooled QC samples may indicate analytical variability and should be filtered out to improve data reliability.

References

1. Wei Q, Zhai X, Song W, Li Z, Pan Y, Li B, Jiao Z, Shi Z, Yu J (2024) Metabolomics analysis of the effects of chelerythrine on Ustilaginoidea virens. J Pestic Sci. https://doi.org/10.1584/jpestics.d23-065
2. Hong L, Wang Y, Zhang Q, Wang Y, Chen M, Li M, Huang Y, Wu Z, Ye J, Wang H (2023) Effects of processing procedures on the formation of aroma intensity and odor characteristic of Benshan tea (Oolong tea, Camellia sentences). Heliyon 9(4):e14855. https://doi.org/10.1016/J.HELIYON.2023.E14855
3. Farag MA, Hegazi N, Dokhalahy E, Khattab AR (2020) Chemometrics based GC-MS aroma profiling for revealing freshness, origin and roasting indices in saffron spice and its adulteration. Food Chem 331. https://doi.org/10.1016/j.foodchem.2020.127358
4. Xu J, Wang M, Zhao J, Wang YH, Tang Q, Khan IA (2018) Yellow tea (Camellia sinensis L.), a promising Chinese tea: processing, chemical constituents and health benefits. Food Res Int 107:567–577. https://doi.org/10.1016/J.FOODRES.2018.01.063
5. Engeseth NJ, Ac Pangan MF (2018) Current context on chocolate flavor development—a review. Curr Opin Food Sci 21:84–91). Elsevier Ltd. https://doi.org/10.1016/j.cofs.2018.07.002
6. Yu P, Yeo ASL, Low MY, Zhou W (2014) Identifying key non-volatile compounds in ready-to-drink green tea and their impact on taste profile. Food Chem 155:9–16. https://doi.org/10.1016/J.FOODCHEM.2014.01.046
7. Zhang W, Bai B, Du H, Hao Q, Zhang L, Chen Z, Mao J, Zhu C, Yan M, Qin H, Abd El-Aty AM (2024) Co-expression of metabolites and sensory attributes through weighted correlation network analysis to explore flavor-contributing factors in various Pyrus spp. Cultivars. Food Chem X 21. https://doi.org/10.1016/j.fochx.2024.101189
8. Wang Y, Kan Z, Thompson HJ, Ling T, Ho CT, Li D, Wan X (2019) Impact of six typical processing methods on the chemical composition of tea leaves using a single Camellia sinensis cultivar, Longjing 43. J Agric Food Chem 67(19):5423–5436. https://doi.org/10.1021/acs.jafc.8b05140
9. Lin Y, Wei R, Zheng JJ, Zheng J, Zhang W, Yasir M, Kayama K, Liu X, Su Z (2024) Volatile and non-volatile compounds profiling and their role in sensory and antioxidative attributes of two species of "red snow tea" (Lethariella). J Food Compos Anal 133:106422. https://doi.org/10.1016/J.JFCA.2024.106422
10. Naveed M, BiBi J, Kamboh AA, Suheryani I, Kakar I, Fazlani SA, FangFang X, Kalhoro SA, Yunjuan L, Kakar MU, Abd El-Hack ME, Noreldin AE, Zhixiang S, LiXia C, XiaoHui Z (2018) Pharmacological values and therapeutic properties of black tea (Camellia sinensis): a comprehensive overview. Biomed Pharmacother 100:521–531. https://doi.org/10.1016/J.BIOPHA.2018.02.048
11. Engelhardt UH (2020) Tea chemistry – what do and what don't we know? – A micro review. Food Res Int 132:109120. https://doi.org/10.1016/J.FOODRES.2020.109120
12. Xing L, Zhang H, Qi R, Tsao R, Mine Y (2019) Recent advances in the understanding of the health benefits and molecular mechanisms associated with green tea polyphenols. J Agr Food Chem 67(4):1029–1043. American Chemical Society. https://doi.org/10.1021/acs.jafc.8b06146
13. Liu L, Wang X, Li Q, Dai H, Zhou Q, Meng X, Chen Z, Zhang X, Zhang Z, Liu Y, Shen T, Yi Y (2024) Metabolomic profiling using UPLC–MS/MS revealed differential changes in metabolites of three tea plant varieties and their potential antioxidant activities. J Food Biochem 2024(1). https://doi.org/10.1155/2024/5566593

14. Yue C, Wang Z, Peng H, Li W, Yang P (2023) UPLC–QTOF/MS-based non-targeted metabolomics coupled with the quality component, QDA, to reveal the taste and metabolite characteristics of six types of Congou black tea. LWT 185:115197. https://doi.org/10.1016/J.LWT.2023.115197
15. Miyauchi S, Yuki T, Fuji H, Kojima K, Yonetani T, Tomio A, Bamba T, Fukusaki E (2014) High-quality green tea leaf production by artificial cultivation under growth chamber conditions considering amino acids profile. J Biosci Bioeng 118(6):710–715. https://doi.org/10.1016/J.JBIOSC.2014.05.008
16. Liu Y, Huang W, Zhang C, Li C, Fang Z, Zeng Z, Hu B, Chen H, Wu W, Wang T, Lan X (2022) Targeted and untargeted metabolomic analyses and biological activity of Tibetan tea. Food Chem 384:132517. https://doi.org/10.1016/J.FOODCHEM.2022.132517
17. Wen M, Zhu M, Han Z, Ho CT, Granato D, Zhang L (2023) Comprehensive applications of metabolomics on tea science and technology: opportunities, hurdles, and perspectives. Compr Rev Food Sci Food Saf 22(6): 4890–4924). John Wiley and Sons Inc. https://doi.org/10.1111/1541-4337.13246
18. Putri SP, Ikram MMM, Sato A, Dahlan HA, Rahmawati D, Ohto Y, Fukusaki E (2022) Application of gas chromatography-mass spectrometry-based metabolomics in food science and technology. J Biosci Bioeng 133(5): 425–435. Elsevier B.V. https://doi.org/10.1016/j.jbiosc.2022.01.011
19. Nolvachai Y, Kulsing C, Marriott PJ (2017) Multidimensional gas chromatography in food analysis. TrAC Trends Anal Chem 96: 124–137). Elsevier B.V. https://doi.org/10.1016/j.trac.2017.05.001
20. Wu W, Zhang L, Zheng X, Huang Q, Farag MA, Zhu R, Zhao C (2022) Emerging applications of metabolomics in food science and future trends. Food Chem X 16. https://doi.org/10.1016/j.fochx.2022.100500
21. Degnan DJ, Bramer LM, Flores JE, Paurus VL, Corilo YE, Clendinen CS (2023) Evaluating retention index score assumptions to refine GC-MS metabolite identification. Anal Chem 95(19):7536–7544. https://doi.org/10.1021/acs.analchem.2c05783
22. Pongsuwan W, Fukusaki E, Bamba T, Yano Y, Yamahara T, Kobayashi A (2006) Prediction of Japanese green tea ranking by gas chromatography/mass spectrometry-based hydrophilic metabolite fingerprinting. J Agric Food Chem 54(19):7444–7449. https://doi.org/10.1021/jf061298v

Chapter 9

GC-MS-Based Metabolite Profiling Method for Sago Starch

Nathanael Steven and Giri Rohmad Barokah

Abstract

Sago starch (*Metroxylon sagu*) is derived from tropical palm tree which widely cultivated across Southeast Asia and Pacific islands. It serves as an important staple crop and is widely recognized for its nutritional and functional properties. However, the biochemical drivers of its characteristic flavor and market quality remain poorly understood. By harnessing the sensitivity and versatility of gas chromatography-mass spectrometry (GC-MS), this chapter introduces a standardized protocol for metabolomic profiling of sago flour. Starting from sample storage, through optimized solvent extraction and derivatization, to GC-MS acquisition and data processing, each step is designed to maximize sensitivity to both volatile and nonvolatile metabolites while ensuring reproducibility and comparability. By providing clear, validated protocols, we aim to enable robust characterization of the metabolites that shape sago starch's taste and aroma, ultimately supporting quality control and product development in the sago industry.

Key words Aroma, Flavor profiles, Metabolomics, Sago starch

1 Introduction

Sago starch (*Metroxylon sagu*), obtained from milling the pith of the sago palm, is an important staple food and carbohydrate source in many regions of Southeast Asia, which include Malaysia, Indonesia, Brunei Darussalam, the Philippines, East Timor, Singapore, and Papua New Guinea [1]. Sago palm has many benefits as a staple food source. First, it is suitable for planting on underutilized land, such as peat swamps and wetlands, where few other staple crops can grow [2]. Its ability to grow in these challenging environments highlights its importance as a sustainable crop for improving food security in the world, providing alternative carbohydrate sources to wheat, rice, corn, and potatoes. Second, sago starch is gluten-free, which is safer for individuals with celiac disease or gluten intolerance. Additionally, almost every part of the plant can be used by humans, such as its leaves, which can be used for roofing, its pith can also be used for stock feed, and its fruit can be made into delicacies [2]. This reinforces sago's value as a sustainable resource.

Sastia Prama Putri (ed.), *Gas Chromatography-Mass Spectrometry: Based Metabolomics in Food Science*, Methods and Protocols in Food Science, https://doi.org/10.1007/978-1-0716-5108-7_9,

Market price and overall quality of sago are mainly influenced by its physicochemical properties and flavor profile [3]. Although Codex Alimentarius has described the general quality standards for edible sago starch [4], the specific metabolites that affect its taste and aroma have yet to be characterized. Traditionally processed sago often develops off-flavors—namely, acidic or bitter notes—and off-aromas such as mustiness or rancidity. These defects are typically associated with the variability in raw material quality, processing techniques, and storage conditions [6–8].

Metabolomics offers a powerful approach for analyzing the complex flavor profiles of foods and pinpointing the compounds that drive their unique tastes and aromas [8]. Among many analytical techniques, GC-MS is known for its exceptional sensitivity to both volatile and nonvolatile metabolites relevant to flavor and aroma [10, 11]. Nonetheless, standardized workflows that link GC-MS metabolite data with sensory quality remain scarce. This chapter fills that gap by detailing optimized GC-MS–based metabolomics protocols for sago starch, designed to deliver robust, reproducible, and comparable results.

2 Materials

2.1 Sample

Sago starch (derived from the pith of *Metroxylon sagu*).

2.2 Non-volatile Compound Analysis

2.2.1 Reagents

1. Methanol (LC-MS grade).
2. Chloroform (LC-MS grade).
3. Ultrapure water.
4. Ribitol (as internal standard).
5. Methoxyamine hydrochloride (98 + %).
6. Ultrapure pyridine (99.5 + %).
7. N-methyl-N-(trimethylsilyl)-trifluoroacetamide (MSTFA).
8. C_8–C_{40} alkane standard mixture.

2.2.2 Laboratory Supplies

1. 50 mL polycarbonate tube.
2. Micropipette.
3. Freezer-safe marker pen for labeling the samples.
4. Freezer gloves for handling frozen objects.
5. −30 °C freezer to store samples.
6. 2.0 mL microcentrifuge tubes.
7. 1.5 mL microcentrifuge tubes.
8. 15 mL centrifuge tubes (for high-volume samples).
9. Vortex mixer.

10. Refrigerated centrifuge.
11. Thermomixer.
12. Vacuum concentrator.
13. GC-MS vials equipped with 200 μL insert.

2.2.3 GC-MS Equipment

1. GC-MS system: GC-MS-QP2010 Ultra (Shimadzu) or equivalent system.
2. GC capillary column: InertCap 5 MS/NP column (GL Sciences) or equivalent column.
3. NIST library (NIST/EPA/NIH EI-MS Library).
4. GL-Science DB (InertCap 5MS Metabolomics, Kovats RI)—can be downloaded on the MS-DIAL website.

2.3 Volatile Compound Analysis (Using Solvent Extraction Method)

2.3.1 Reagents

1. Dichloromethane (≥99.8%).
2. N-alkenes (C_7–C_{32}).
3. Pure nitrogen (N_2) gas.

2.3.2 Laboratory Supplies

1. Screw bottle 50 mL for sample extraction.
2. Magnetic stirrer for sample extraction.
3. Hotplate stirrer for sample extraction.
4. Micropipette.
5. Freezer-safe marker pen for labeling the samples.
6. Freezer gloves for handling frozen objects.
7. −30 °C Freezer to store samples.
8. GC-MS vials.
9. Filter paper (size 110 mm).
10. Spirit tube (5 mL).
11. Nitrogen concentrator.

2.3.3 GC-MS Equipment

1. GC-MS system: GC/MS-TQ8040 system (Shimadzu Co., Kyoto, Japan) or equivalent system.
2. GC capillary column: InertCAP FFAP column (60 m × 0.32 mm I.D., df. = 0.50 mm) (GL Sciences Inc., Tokyo, Japan).
3. NIST library (NIST/EPA/NIH EI-MS Library).

3 Methods

Sago starch consists of roughly 70–80% amylopectin and 20–30% amylose [11]. Its high macromolecular content can cause sample "stickiness" within the GC-MS column, leading to carryover and degraded performance. To prevent adhesion and preserve column integrity, increased pyridine injections as the solvent in between samples are effective to reduce stickiness without compromising detection.

3.1 Sample Preparation

Because sago starch is supplied as a fine flour, no pretreatment is required prior to extraction. To preserve its chemical composition, samples should be stored in airtight, light-protected containers at cool temperatures, avoiding direct sunlight.

3.2 Nonvolatile Compound Analysis

3.2.1 Sample Extraction

1. Prepare the extraction solvent consisting of methanol, chloroform, and ultrapure water in a ratio of 2.5:1:1 (v/v/v) and an internal standard (ribitol 0.05 mg/mL) (*see* **Note 1**).
2. Place 20 milligrams ($n = 3$) of the homogenized sample powder in a 2 mL microcentrifuge tube (*see* **Note 2**).
3. Add 2000 μL of extraction solvent into a 2 mL microcentrifuge tube. Prepare microcentrifuge tubes for blank samples as well. These blank samples of microcentrifuge tubes contain only extraction solvent (no starch samples).
4. Vortex and incubate all samples in a thermomixer (1200 rpm, 37 °C for 30 min).
5. Centrifuge for 3 min at 4 °C, 10,000 rpm.
6. Collect 800 μL of the supernatants and transfer to a new 2 mL microcentrifuge tube.
7. Add 800 μL of ultrapure water and vortex.
8. Centrifuge samples again for 3 min at 4 °C, 10,000 rpm.
9. Separate supernatants into two parts. Transfer 1000 μL of supernatant to a new 1.5 mL microcentrifuge tube. Collect 400 μL of supernatant in each sample for quality control (QC) in a 15 mL centrifuge tube, then vortex.
10. Transfer 600 μL of mixed supernatant from all samples into a new 1.5 mL microcentrifuge tube for making QC samples.
11. Evaporate all the samples to dryness with a vacuum concentrator for 2 hours at room temperature (±25 °C)—until no water is left (*see* **Note 3**).
12. Store the samples in a −30 °C freezer till analysis.

3.2.2 Derivatization

1. Thaw frozen samples at room temperature (±25 °C).
2. Freshly prepare methoxyamine hydrochloride (20 mg/mL) in pyridine and mix vigorously until dissolved.
3. Add 100 μL of methoxyamine hydrochloride solvent to the dried samples, vortex, and flash down the liquid.
4. Incubate the samples in a thermomixer at 30 °C for 90 min and 1200 rpm.
5. Add 50 μL of MSTFA to each sample.
6. Incubate the resulting solution at 37 °C for 30 minutes and 1200 rpm.
7. Transfer 100 μL of reaction solution into a GC-MS vial and cap firmly. Samples are ready to inject.

3.2.3 GC-MS Parameter

1. The injection amount is one μL with a split ratio of 1:25.
2. The injection temperature is set to 270 °C.
3. Set the hydrogen carrier gas flow rate to 1.12 mL/min.
4. The column temperature gradient was initially set at 80 °C for 2 minutes, then increased by 15 °C/min until reaching the final temperature of 330 °C—hold for 6 minutes.
5. The transfer line temperature is set to be maintained at 310 °C, while the ion source temperature is set at 280 °C.
6. The electron ionization source operates at −70 eV.
7. The mass spectrometer operates in scan mode at a rate of 10,000 μ/s within a mass range of 85–500 *m/z*.
8. A standard alkane mixture (C_8–C_{40}) is set to be injected first before the samples.

3.3 Volatile Compound Analysis

3.3.1 Sample Extraction

1. Weigh 10 g of sago starch samples in the 50 mL screw bottle.
2. Add 30 mL of dichloromethane to the sample.
3. Place the magnetic stirrer in the mixture of the sample and dichloromethane.
4. Stir the samples using a magnetic stirrer at 100 rpm for 12 hours.
5. Collect the component of dichloromethane in a flask and transfer it into a spirit tube.
6. Concentrate the sample using a nitrogen concentrator into 200 μL.
7. The samples obtained were then transferred to a GC/MS vial and stored in a freezer at −30 °C until further analysis.

3.3.2 GC-MS Parameter

1. The analyte was injected at a split ratio of 10:1 (v/v).
2. The injection section temperature is set at 250 °C.
3. Hydrogen is used as the carrier gas.
4. The linear velocity and purge flow rates were set at 40.7 cm/s and 2.5 mL/min.
5. After the sample injection, the column oven was maintained at 40 °C for 5 min.
6. The temperature was subsequently raised to 250 °C at a heating rate of 3 °C and then maintained at 250 °C for 45 min.
7. MS was operated in the scan mode (m/z 24–350) with the ion source and interface temperature set at 250 °C.
8. A standard mixture of alkanes (C_7-C_{32}) was used to calculate the retention indices.

4 Notes

1. It is recommended that the extraction solvent be prepared freshly to avoid oxidation of chloroform.
2. Weigh samples in a controlled, air-conditioned environment (20–25 °C) while wearing powder-free gloves to avoid contamination and moisture uptake. Because sago flour readily clings under electrostatic charge, antistatic measures such as using ionizing air blowers are necessary to ensure accurate, reproducible mass measurements.
3. Maintain the tubes at around 25 °C and confirm they are completely free of water. An extended drying period is possible if any residual solvent remains. This ensures sample dryness and minimizes matrix effects.

References

1. Pendi FH, Yan W-J, Hussain H, Roslan HA, Julaihi N (2023) Advances in sago palm research: a comprehensive review of recent findings. Sains Malaysiana 52(11): 3045–3059. https://doi.org/10.17576/jsm-2023-5211-03
2. Ehara H (2018) Genetic variation and agronomic features of Metroxylon palms in Asia and Pacific. In: Ehara H, Toyoda Y, Johnson DV (eds) Sago palm: multiple contributions to food security and sustainable livelihoods. Springer Nature, pp 45–60. https://doi.org/10.1007/978-981-10-5269-9
3. Trisia MA, Takeshita H, Kikuta M, Ehara H (2020) Factors determining sago starch import demand: empirical evidence from Japan. J Int Logist Trade 18(4):159–167. https://doi.org/10.24006/JILT.2020.18.4.159
4. Codex Alimentarius Commission (2011) Regional standard for edible sago flour (CXS 301R-2011; amended in 2023). Food and Agriculture Organization (FAO) & World Health Organization (WHO). https://www.fao.org/fao-who-codexalimentarius
5. Diyono W, Budiyanto A, Purwani EY, Aulia S, Rumaisha, Widaningrum (2022) Physical properties of physico-enzymatically modified sago starch originated from Papua: a preliminary study. IOP Conf Ser Earth Environ Sci 1024(1):012045. https://doi.org/10.1088/1755-1315/1024/1/012045

6. Pinyo J, Luangpituksa P, Suphantharika M, Hansawasdi C, Wongsagonsup R (2017) Improvement of sago starch extraction process using various pretreatment techniques and their pretreatment combination. Starch/Stärke 69(9–10):1700005. https://doi.org/10.1002/star.201700005
7. Greenhill AR, Blaney BJ, Shipton WA, Frisvad JC, Pue A, Warner JM (2008) Mycotoxins and toxigenic fungi in sago starch from Papua New Guinea. Lett Appl Microbiol 47(4):342–347. https://doi.org/10.1111/j.1472-765X.2008.02439.x
8. Diez-Simón C, Mumm R, Hall RD (2019) Mass spectrometry–based metabolomics of volatiles as a new tool for understanding aroma and flavour chemistry in processed food products. Metabolomics 15(3):41. https://doi.org/10.1007/s11306-019-1493-6
9. Kitani Y, Putri SP, Fukusaki E (2022) Investigation of the effect of processing on the component changes of single-origin chocolate during the bean-to-bar process. J Biosci Bioeng 134(2):138–143. https://doi.org/10.1016/j.jbiosc.2022.05.007
10. Bastos VS, Uekane TM, Bello NA, de Rezende CM, Flosi Paschoalin VM, Del Aguila EM (2019) Dynamics of volatile compounds in TSH 565 cocoa clone fermentation and their role on chocolate flavor in Southeast Brazil. J Food Sci Technol 56(6):2874–2887. https://doi.org/10.1007/s13197-019-03736-3
11. Yamazaki T, Honda Y, Nishiyama S, Okazaki M (2017) Difference in the amylopectin chain distribution of sago (*Metroxylon sagu*) starch from Papua New Guinea and The Philippines. Sago Palm 25:1–5

Chapter 10

GC-MS-Based Metabolite Profiling Method for Tropical Fruits

Muhammad Maulana Malikul Ikram

Abstract

Tropical fruits are essential for the macronutrient and micronutrient needs of a person. A comprehensive understanding of the nutrients or compounds from tropical fruits could elucidate the benefits more deeply and unlock potential for future health. Gas chromatography-mass spectrometer (GC-MS) analysis is an important instrument for analyzing comprehensive metabolites, also known as metabolomics, from samples, including tropical fruits. However, several metabolites need to be separated and also derivatized to ensure their ability to be detected by GC-MS. Therefore, proper extraction and derivatization methods are crucial to give clarity to the produced result. This book chapter explains about the method used in several research paper of GC-MS tropical fruits analysis that use Bligh-Dyer method-base extraction and double derivatization steps from oximation and silylation. Using a combination of these steps, these research can detect more than 50 metabolites from several classes, such as amino acids, sugars, sugar alcohols, organic acids, etc. Therefore, the detailed materials, methods, and notes acquired from experience are hoped to help other tropical fruit researchers to perform GC-MS analysis in the future.

Key words Extraction, Derivatization, Gas chromatography-mass spectrometer, Tropical fruits, Metabolomics

1 Introduction

The study of tropical fruits using gas chromatography-mass spectrometry (GC/MS) is critical for understanding their complex chemical composition, which is required for applications such as food quality control, nutritional analysis, and bioactive ingredient identification. Tropical fruits contain a variety of metabolites, including sugars, organic acids, fatty acids, and phytochemicals. These compounds need to be treated with proper extraction and derivatization prior to GC-MS analysis. The extraction process separates the target metabolite from the complex matrices, while derivatization converts the metabolites from nonvolatile to volatile due to increased thermal stability and volatility. The combination of

Sastia Prama Putri (ed.), *Gas Chromatography-Mass Spectrometry: Based Metabolomics in Food Science*, Methods and Protocols in Food Science, https://doi.org/10.1007/978-1-0716-5108-7_10,

extraction and derivatization processes enables comprehensive metabolite analysis, or metabolomics, of tropical fresh fruits.

The extraction procedure is an important step in preparing samples for GC-MS analysis, especially for complex matrices such as tropical fruits. Tropical fruits include a wide range of volatile and nonvolatile substances, including sugars, organic acids, fatty acids, and phytochemicals like terpenes and phenols. The extraction procedure can isolate target compounds from matrix interference, thus increasing the accuracy that is crucial for assessing food quality, flavor profiling, and detection of bioactive compounds. Therefore, selecting the appropriate extraction method is important for the clarity of GC-MS tropical fruit analysis [1]. The Bligh-Dyer method, developed in 1959, is an extraction technique to analyze food matrices based on their hydrophilic and hydrophobic compounds. The extraction procedure begins with mixing the sample with a methanol-chloroform-water mixture, then adding water to disrupt the solution's equilibrium, thus separating the content. For low-molecular-weight metabolomics investigation with GC-MS, the top layer containing lipophilic compounds is collected and studied further. This method is highly useful for extracting hydrophilic compounds such as sugars, amino acids, and organic acids, making it a useful tool in GC/MS processes for tropical fruits [2].

The Bligh-Dyer approach has the benefit of remarkable efficacy in separating lipids from biological matrices, even at low lipid levels. Furthermore, the procedure is simple and does not require costly equipment, making it accessible to many laboratories. However, the use of chloroform, a toxic and environmentally dangerous solvent, presents substantial safety and disposal problems. Moreover, the approach may not be suitable for extracting highly polar or heat-sensitive molecules, limiting its usefulness for comprehensive metabolomic research. As a result, while the Bligh-Dyer method remains a cornerstone in lipidomics, researchers frequently alter or combine it with other approaches to address these shortcomings and expand its scope [3].

Regarding the derivatization process, oximation is a derivatization technique widely used in GC/MS analysis to improve the detectability and stability of chemicals containing carbonyl groups, such as sugars. The interaction of carbonyl groups in sugars with methoxyamine hydrochloride produces oximes that make the sugars more volatile and detectable using GC-MS. Typically, the derivatization process begins with dissolving the sample extract in a suitable solvent, such as pyridine, and then adding methoxyamine hydrochloride. To achieve complete conversion of carbonyl groups, the reaction is carried out at a certain temperature, typically 60 °C, for a set period of time. Oximation not only improves the volatility and chromatographic characteristics of analytes but also prevents their breakdown during the high-temperature conditions of GC/MS analysis [4]. While oximation is for carbonyl-containing groups, silylation is a popular technique for derivatizing compounds with hydroxyl, carboxyl, or amine functional groups. *N*-methyl-*N*-(trimethylsilyl)trifluoroacetamide (MSTFA) is a popular silylating reagent due to its effectiveness and versatility. This procedure substitutes active hydrogens in functional groups with trimethylsilyl groups, which improves the analytes' thermal stability and volatility, thus allowing the compounds to be detected in GC/MS. Silylation with MSTFA is useful for studying polar metabolites such as organic acids, amino acids, and carbohydrates that have a great impact on tropical fruit acceptability [4].

Combining oximation and silylation methods for derivatization has substantial advantages for GC/MS analysis, especially for complex matrices like tropical fruits. This dual-derivatization method enables complete profiling of a diverse variety of metabolites. The oximation procedure stabilizes carbonyl groups, preventing degradation and improving chromatographic performance, while silylation enhances the volatility and thermal stability of hydroxyl, carboxyl, and amine functional groups. Thus, the combination enables a thorough depiction of the metabolome by addressing the derivatization needs of various chemical groups [5]. However, this method lengthens sample preparation due to the necessity of sequential application between two derivatization procedures. Despite these challenges, the combination of oximation and silylation has proven to be an effective technique for metabolomic research on tropical fruits.

2 Materials

2.1 Sample Preparation

Liquid nitrogen.

2.2 Extraction

1. Microtube 2 mL and 1.5 mL.
2. Holed-microtube caps.
3. Extraction solvent (methanol:chloroform:ultrapure water) (5: 2:2 ratio v/v/v) (*see* **Notes 1–4**).
4. Ribitol dissolved in ultrapure water (100 μg/mL) (*see* **Note 3**).

2.3 Derivatization

1. Methoxyamine hydrochloride in pyridine (20 mg/mL).
2. N-methyl-N-(trimethylsilyl)trifluoroacetamide (MSTFA) ampule.

2.4 Instrument

1. Multi-beads shocker with metal beads (*see* **Note 5**).
2. Weighing machine.
3. Micropipette.
4. Vortex machine.

5. Thermomixer incubator with shaker mechanism.
6. Refrigerated centrifuge.
7. Centrifugal evaporator.
8. Freeze dryer.

3 Methods

These extraction and derivatization methods were already used in several research papers that used tropical fruits as the sample, such as pineapple, mangosteen, banana, and mango [6–11]; the detailed method was described below.

3.1 Sample Preparation

1. Separate fruit sample between its flesh, peel, and other parts (leaf, crown, etc.) (*see* **Note 6**).
2. Cut the flesh fruit into 1 × 1 × 1 cm or 2 × 2 × 2 cm and quench with liquid nitrogen and freeze dry overnight (*see* **Note 7**).
3. Cut the peel and leaf part into 1 × 1 cm or 2 × 2 cm and quench with liquid nitrogen and freeze dry overnight (*see* **Note 7**).
4. Homogenize all the freeze-dried samples with multi-beads shocker using metal beads until the fruit sample becomes powder and keep in −80 °C refrigerator before use for extraction (*see* **Notes 5**, **8**, **9**).
5. For a sample containing water, such as coconut, 100 μL of fruit water was directly taken into the microtube 2 mL (*see* **Note 10**).

3.2 Extraction

1. Fruit samples were weighed until reaching 5−10 mg using a weighing machine in a 2 mL microtube (*see* **Notes 10** and **11**).
2. The sample was added with 1 mL of mixed solvent using a micropipette and then vortexed to ensure homogeneity (*see* **Note 12**).
3. Samples were incubated in a thermomixer incubator with the condition of 37 °C, 1200 rpm for 30 min (*see* **Note 13**).
4. Incubated samples were centrifuged for 3 min at 4 °C, 10,000 rpm for separating the solid and liquid phase (*see* **Note 13**).
5. Take 600 μL of supernatant and transfer to a new 1.5 mL microtube, and add 300 μL of ultrapure water.
6. Re-centrifuge the sample for 3 min at 4 °C, homogenizing at 10,000 rpm for separating the hydrophilic and hydrophobic phase (*see* **Note 13**).

7. Take 400 μL of supernatant (hydrophilic phase) and transfer to a new 1.5 mL microtube and close with holed caps.
8. Put the microtube into the vacuum evaporator to evaporate the solvent for 1 h at room temperature (*see* **Notes 13** and **14**).
9. After only a small amount of solvent is present, lyophilize the sample using a freeze-dry instrument overnight, with the condition of the freeze-dry machine is around 10 Pa below 5 °C.

3.3 Derivatization

1. Add methoxyamine hydrochloride 20 mg/mL in pyridine to the lyophilized sample for 100 μL and close again with normal caps.
2. Incubate the first-derivatized sample in the thermomixer for 90 min at 30 °C, 1200 rpm.
3. Add 50 μL of MSTFA to the sample (*see* **Note 15**).
4. Incubate the second-derivatized sample in the thermomixer for 30 min at 37 °C, 1200 rpm.
5. Centrifuge the sample for 3 min at 30 °C, 10,000 rpm for separating the debris from the supernatant (*see* **Note 16**).
6. Take 100 μL of the derivatized sample to the GC/MS vial.

4 Notes

1. Prepared the mixed solvent stock prior to analysis for easier nine-multiplication volume (following 5:2:2 ratio). For example, prepare the mixed solvent stock in 36 mL for 30 samples, 72 mL, and so on.
2. Mixed solvent that has been kept for more than 1 month is better replaced with a new one.
3. Ribitol used as an internal standard needs to be optimized based on the sample. The choice of ribitol needs to be based on the group similarity with the samples, but it was not present in the sample. For example, ribitol is a group of sugar alcohols that are commonly found in fruit samples. However, ribitol is not found naturally in the sample. Therefore, it was chosen as the internal standard for the fruit sample.
4. Mixed solvents need to be prepared in order of hydrophilic to hydrophobic (ultrapure water > methanol > ribitol > chloroform) for homogeneity.
5. Alternative for researchers who do not own a multibeads shocker is a mortar-pestle with liquid nitrogen.
6. Separate the fruits between their flesh, peel, and other parts of the fruit for better homogeneity and clearer results (unbiased due to differences in fruit parts).

7. Fruit samples that have newly arrived should be directly quenched with liquid nitrogen and kept at −80 °C refrigerator to ensure the metabolites are kept fresh and represent the current condition.
8. Flesh and peel samples can achieve homogeneity using a multi-beads shocker with one cycle of crushing, while leafy parts usually take two cycles of crushing due to the high concentration of fiber.
9. Utilization of the multi-beads shocker was more prominent compared to the food processor or juicer due to its ability to crush the sample under cold conditions, thus preventing deterioration of metabolites caused by heat during the homogenization process.
10. Fruit samples in the form of a liquid can be directly added to a mixed solvent for liquid-liquid extraction.
11. Fruit sample weight for GC/MS analysis can be optimized between 5 mg and 10 mg. For sugar analysis, usually the smaller the amount, the better for keeping the GC/MS working in good condition. If we remove the sugar from GC/MS analysis (whether through SPE or filament-off analysis during sugar peak), 10 mg would be preferable to enhance other metabolite peak detection.
12. Fruit samples are high in sugar content; therefore, several researchers use solid-phase extraction (SPE) to remove the sugar content during GC/MS analysis.
13. Put the samples in the balance area while using the thermomixer, and centrifuge to ensure the same forces are applied to the sample.
14. During the evaporator centrifuge steps, ensure the alcohol in the machine is filled until it submerges in the coil. If not, the sample will not be fully evaporated and will hinder the derivatization steps.
15. During the derivatization process, a new ampule of MSTFA will give better results compared to the old/opened one.
16. GC/MS instrument is a sensitive instrument; therefore, it will be better to centrifuge after the silylation step to ensure the sample that will be moved to the GC/MS vial is cleaned of any debris.

References

1. Hurkul MM, Cetinkaya A, Yayla S, Ozkan SA (2024) Advanced sample preparation and chromatographic techniques for analyzing plant-based bioactive chemicals in nutraceuticals. J Chromatogr Open 5:100131. https://doi.org/10.1016/j.jcoa.2024.100131
2. Bligh EG, Dyer WJ (1959) A rapid method of total lipid extraction and purification. Can J

Biochem Physiol 37(8):911–917. https://doi.org/10.1139/o59-099

3. Iverson SJ, Lang SL, Cooper MH (2001) Comparison of the Bligh and Dyer and Folch methods for total lipid determination in a broad range of marine tissue. Lipids 36(11):1283–1287. https://doi.org/10.1007/s11745-001-0843-0
4. Villas-Bôas SG, Smart KF, Sivakumaran S, Lane GA (2011) Alkylation or silylation for analysis of amino and non-amino organic acids by GC-MS? Meta 1(1):3–20. https://doi.org/10.3390/metabo1010003
5. Miyagawa H, Bamba T (2019) Comparison of sequential derivatization with concurrent methods for GC/MS-based metabolomics. J Biosci Bioeng 127(2):160–168. https://doi.org/10.1016/j.jbiosc.2018.07.015
6. Ikram MMM, Mizuno R, Putri SP, Fukusaki E (2021) Comparative metabolomics and sensory evaluation of pineapple (Ananas comosus) reveal the importance of ripening stage compared to cultivar. J Biosci Bioeng 132(6):592–598. https://doi.org/10.1016/j.jbiosc.2021.08.008
7. Ikram MMM, Putri SP, Fukusaki E (2023) Chitosan-based coating enriched with melezitose alters primary metabolites in fresh-cut pineapple during storage. J Biosci Bioeng 136(5):374–382. https://doi.org/10.1016/j.jbiosc.2023.08.002
8. Ikram MMM, Ridwani S, Putri SP, Fukusaki E (2020) GC-MS-based metabolite profiling to monitor ripening-specific metabolites in pineapple (Ananas comosus). Meta 10(4):134. https://doi.org/10.3390/metabo10040134
9. Parijadi AAR, Yamamoto K, Ikram MMM, Dwivany FM, Wikantika K, Putri SP, Fukusaki E (2022) Metabolome analysis of banana (Musa acuminata) treated with chitosan coating and low temperature reveals different mechanisms modulating delayed ripening. Front Sustain Food Syst 6:835978. https://doi.org/10.3389/fsufs.2022.835978
10. Rahmawati D, Adan MFY, Ikram MMM, Iman MN, Fukusaki E, Putri SP (2024) Effect of sodium metabisulfite treatment and storage condition on metabolic profile of young coconut (Cocos nucifera L.). J Biosci Bioeng 138(6):515–521. https://doi.org/10.1016/j.jbiosc.2024.08.002
11. Sato M, Ikram MMM, Pranamuda H, Agusta W, Putri SP, Fukusaki E (2021) Characterization of five Indonesian mangoes using gas chromatography-mass spectrometry-based metabolic profiling and sensory evaluation. J Biosci Bioeng 132(6):613–620. https://doi.org/10.1016/j.jbiosc.2021.09.006

Chapter 11

GC-MS-Based Metabolite Profiling Method for Dried Mango

Waqif Agusta

Abstract

This chapter outlines a protocol for extracting and derivatizing polar metabolites in dried mango (*Mangifera indica*) for gas chromatography-mass spectrometry (GC-MS) analysis. Due to their high sugar content and complex matrix, dried mangoes present challenges for polar compound profiling, necessitating a robust sample preparation approach. The method differentiates between sugars (e.g., glucose, fructose, sucrose) and non-sugar compounds (e.g., amino acids and organic acids) by employing separate extraction pathways. Cryogenically ground samples are extracted using mixed solvents (methanol:chloroform:water, 5:2:2, v/v/v) with ribitol as an internal standard. Extracts are phase-separated, and aliquots are taken for sugar and non-sugar analysis. Derivatization involves a sequential oximation step using methoxyamine hydrochloride, followed by silylation with MSTFA, enabling improved volatility and thermal stability of analytes. The resulting GC-MS workflow facilitates comprehensive metabolite profiling for nutritional analysis, process monitoring, and product quality control in dried mango and other dehydrated tropical fruits.

Key words Extraction, Derivatization, Gas chromatography-mass spectrometry, Dried mango, Metabolomics

1 Introduction

Mango (*Mangifera indica*) is one of the most widely cultivated and consumed tropical fruits globally. In addition to its vibrant color, sweetness, and aromatic complexity, mango offers a rich nutritional profile, containing abundant carbohydrates, vitamins, polyphenols, and organic acids [1, 2]. These properties contribute to its popularity as a fresh fruit and a processed product. However, the high moisture content and rapid postharvest ripening of fresh mango limit its shelf life and commercial transportability. To address these constraints, drying and dehydration techniques have become preferred for preserving mangoes, producing shelf-stable forms such as dried mango slices and mango leather [3]. Drying is a widely used preservation method for fruits, extending shelf life and reducing postharvest losses [4–7]. This technique decreases moisture content, minimizing bulk weight and facilitating year-round availability

Sastia Prama Putri (ed.), *Gas Chromatography-Mass Spectrometry: Based Metabolomics in Food Science*, Methods and Protocols in Food Science, https://doi.org/10.1007/978-1-0716-5108-7_11,

[6]. Dried fruits are nutrient-dense, containing concentrated vitamins, minerals, and bioactive compounds. The dried products find applications in various industries, including confectionery, bakery, and food supplements [4]. Overall, drying technologies provide cost-effective solutions for preserving fruits and vegetables while maintaining nutritional value [4, 5].

During dehydration, mangoes undergo substantial physicochemical and biochemical changes, including water removal, concentration of sugars and organic acids, and partial degradation of heat-labile vitamins [8]. These transformations affect the dried product's nutritional composition and sensory characteristics, including flavor, color, and texture. Understanding the chemical composition of dried mango, particularly its polar metabolite profile, is essential for ensuring quality, detecting adulteration, optimizing drying parameters, and meeting regulatory standards.

Among the many constituents present in dried mango, polar metabolites such as mono- and disaccharides (glucose, fructose, sucrose), sugar alcohols (sorbitol, mannitol), organic acids (citric, malic, succinic), and amino acids (glutamic acid, alanine, aspartic acid) are of analytical interest. These compounds contribute to the product's taste profile (sweetness, acidity, and umami), preservation properties, and nutritional value. Moreover, specific metabolite ratios may indicate processing conditions, varietal authenticity, or ripening stage before drying. As such, metabolite fingerprinting of dried mango has essential applications in food authentication, quality control, and product development.

Gas chromatography-mass spectrometry (GC-MS) has emerged as a powerful analytical tool for comprehensively profiling these polar compounds. GC-MS provides high-resolution separation and mass-based identification of low-molecular-weight compounds, making it especially suitable for metabolomics studies in complex food matrices such as dried fruits [9]. However, the successful application of GC-MS to dried mango analysis presents several analytical challenges. Most polar metabolites are nonvolatile, thermally unstable, or prone to decomposition under the high temperatures used in GC injection. Therefore, effective sample extraction and chemical derivatization are critical prerequisites for enabling GC-MS detection.

The extraction process is crucial in releasing analytes from the fruit matrix and minimizing interference from nontarget macromolecules such as starches, pectins, and proteins. Dried mango tissue presents a dense hygroscopic matrix that hinders efficient solvent penetration. Furthermore, the high sugar content increases viscosity and may trap other compounds, selecting solvent system, temperature, and agitation parameters crucial for the successful recovery of hydrophilic compounds. A widely adopted protocol in food and plant metabolomics is the Bligh and Dyer method, which utilizes a ternary solvent system—methanol, chloroform, and water—to partition metabolites into hydrophilic and hydrophobic fractions [10]. Although initially developed for lipid analysis, this method has proven effective for recovering polar metabolites from plant-based matrices, including fruits [11]. The aqueous phase obtained from this extraction is particularly suitable for analyzing sugars, amino acids, and organic acids.

While the Bligh-Dyer method offers high extraction efficiency and simplicity, it has limitations. Using chloroform, a toxic and environmentally hazardous solvent, raises safety concerns and complicates waste disposal. Additionally, extremely polar or heat-sensitive metabolites may be incompletely extracted or degraded during the process. Consequently, several studies have introduced modifications to the solvent ratios or incorporated additional clean-up steps to improve compatibility with GC-MS workflows [12]. Nonetheless, the method remains one of the most accessible and reproducible extraction protocols for dried fruit matrices and is widely employed in metabolomic investigations of mango and related species.

Following extraction, chemical derivatization is required to modify analytes for GC-MS analysis. Derivatization improves the thermal stability, volatility, and chromatographic behavior of compounds that would otherwise be unsuitable for gas-phase separation. In dried mango profiling, two complementary derivatization techniques are commonly used: oximation and silylation.

Oximation targets carbonyl-containing metabolites, particularly reducing sugars such as fructose and glucose, which tend to degrade or isomerize under GC-MS conditions. This reaction is achieved by incubating the sample extract with methoxyamine hydrochloride in pyridine, typically at 60 °C for 60–90 min. The resulting oxime derivatives exhibit improved thermal stability and peak symmetry, allowing for more reliable detection and quantification [13]. Oximation also minimizes the formation of multiple tautomers during GC separation, thus simplifying spectral interpretation.

After oximation, the samples undergo silylation, in which functional groups such as hydroxyls, carboxyls, and amines are derivatized with reagents such as N-methyl-N-(trimethylsilyl)-trifluoroacetamide (MSTFA). This process replaces active hydrogen atoms with trimethylsilyl (TMS) groups, converting polar molecules into more volatile, GC-amenable forms. Silylation is particularly useful for derivatizing amino acids, sugar alcohols, and organic acids, compounds abundant in dried mango and contributing to flavor and nutritional quality [14].

Combining oximation and silylation provides broad compound coverage and enhances data quality by stabilizing sensitive functional groups and improving detection sensitivity [15]. Despite extending the sample preparation time, this dual derivatization strategy is considered standard in GC-MS-based metabolomics and widely used to analyze complex fruit-derived matrices. For dried mango, this approach enables high-resolution profiling of compounds that are otherwise difficult to detect, such as minor amino acids, fermentation byproducts, and degradation products of sugars.

Recent studies have demonstrated the utility of GC-MS in assessing dried mango quality by monitoring changes in polar metabolite composition during different drying treatments. GC-MS analysis has been used to profile volatile, nonpolar, and polar metabolites in various mango cultivars, revealing significant intervarietal differences [16]. This technique has also been employed to evaluate metabolic perturbations in mangoes ripened using different agents, identifying 16 differential metabolites and four altered metabolic pathways [17]. Dried mango is a valuable model for GC-MS-based polar metabolite profiling due to its nutritional richness, processing complexity, and commercial importance. A robust analytical workflow combining optimized extraction with oximation–silylation derivatization allows for comprehensive characterization of sugars, amino acids, organic acids, and related compounds. These data can be used to support quality assurance, authenticity testing, process optimization, and functional ingredient discovery. The following sections of this chapter describe the methodologies employed for extracting and deriving polar compounds in dried mango, providing a reproducible framework for GC-MS-based metabolomics in fruit systems.

2 Materials

2.1 Sample Preparation

Liquid nitrogen.

2.2 Extraction

1. Microtube 2 mL and 1.5 mL.
2. Holed-microtube caps.
3. Mixed solvent (methanol: chloroform: ultrapure water) (5,2,2, ratio v/v/v) (*see* **Notes 1 and 2**).
4. Ribitol dissolved in ultrapure water (100 μg/mL) (*see* **Note 3 and 4**).

2.3 Derivatization

1. Methoxyamine hydrochloride in pyridine (20 mg/mL).
2. N-methyl-N-(trimethylsilyl)trifluoroacetamide (MSTFA) ampule.

2.4 Instrument

1. Multi-beads shocker with metal beads.
2. Weighing machine.
3. Micropipette.

4. Vortex machine.
5. Thermomixer incubator with shaker mechanism.
6. Refrigerated centrifuge.
7. Centrifugal evaporator.
8. Freeze dryer.

3 Methods

A structured extraction procedure was implemented to enable accurate GC-MS profiling of polar compounds in dried tropical fruits. Given the wide dynamic range and diverse abundances of target metabolites, the protocol was divided into two analytical workflows:

(a) Sugar compounds (e.g., fructose, sorbose, sucrose).
(b) Non-sugar hydrophilic metabolites (e.g., amino acids, organic acids, sugar alcohols).

This separation is necessary due to differences in compound abundance and analytical requirements. Sugars, typically in high concentrations in dried fruits, require lower sample input to prevent detector saturation and minimize ion suppression effects. In contrast, non-sugar metabolites tend to occur at lower concentrations and thus benefit from increased sample mass to enhance detection sensitivity. This stratified approach improves quantification accuracy, chromatographic resolution, and overall reproducibility.

3.1 Sample Preparation

1. Dried fruit samples were finely chopped and cryogenically frozen using liquid nitrogen and freeze-dried overnight (*see* **Note 5**).
2. Samples were then pulverized into a fine powder using a multi-beads shocker (*see* **Notes 6 and 7**).

3.2 Extraction

3.2.1 Extraction Procedure

1. Perform sugar and non-sugar compounds analyses separately using the same procedure. Weigh 5 mg of powdered sample for sugar compounds analysis and 10 mg for non-sugar compounds analysis into a 2 mL microtube (*see* **Note 8**).
2. Add 1 mL of mixed solvent containing ribitol (100 μg/mL) as an internal standard using a micropipette, and then vortex the sample to ensure homogeneity (*see* **Note 9**).
3. Prepare procedure blanks with solvent only.
4. Incubate in the Thermo-shaker incubator at 37 °C, 1200 rpm for 30 min. (*see* **Note 10**).

5. Vortex and centrifuge the samples for 3 min at 4 °C, 10,000 rpm for separating the solid and liquid phase (*see* **Note 10**).
6. Transfer 600 μL of supernatant to a new 1.5 mL microtube.
7. Add 300 μL of ultrapure water, vortex, and centrifuge the sample for 3 min at 4 °C, 10,000 rpm to homogenize and separate the hydrophilic and hydrophobic phases (*see* **Note 10**).
8. Separate the supernatants into two parts. Transfer 50 μL of supernatant to a new 1.5 mL microtube, collect another 50 μL of supernatant, and gather it in the QC pool (2 mL microtube).
9. Transfer 50 μL of mixed supernatant from all samples (QC pool) into a new 1.5 mL microtube to make QC samples.
10. Close all the samples with holed caps and put the microtube into the vacuum evaporator to evaporate the solvent for 1 h at room temperature, until no water is left (*see* **Notes 10** and **11**).
11. Store extracts at 4 °C until derivatization (*see* **Note 12**).

3.3 Derivatization

1. Methoxyamine hydrochloride 20 mg/mL in pyridine was added to the lyophilized sample for 100 μL and closed again with standard caps. (*see* **Note 13**).
2. Incubate the first-derivatized sample in a Thermo-shaker for 90 min at 30 °C, 1200 rpm.
3. Add 50 μL of MSTFA to the sample (*see* **Note 14**).
4. Incubate the second-derivatized sample in a Thermo-shaker for 30 min at 37 °C, 1200 rpm.
5. Centrifuge the sample for 3 min at 25 °C, 10,000 rpm for separating the debris from the supernatant (*see* **Note 15**).
6. Took 100 μL of the derivatized sample to the GC/MS vial.

3.4 Gas Chromatography-Mass Spectrometry Analysis

1. Instrument Preparation
   (a) Use either GCMS-QP2010 Ultra or GCMS-TQ8030 (Shimadzu, Kyoto, Japan) equipped with an autosampler AOC-20i/s (Shimadzu, Kyoto, Japan).
   (b) Install a guard column: InertCap 5MS/NP ProGuard, 35 m × 0.18 mm I.D., 0.18 μm df; GL Sciences).
2. Tuning and Calibration
   (a) Tune and calibrate the mass spectrometer before sample analysis.
   (b) Inject a standard alkane mixture (C_9-C_{40}) to enable calculation of linear retention indices (RI).

3. GC-MS Settings
   (a) Set the injection mode to split (25,1, v/v).
   (b) Set the injection temperature to 270 °C.
   (c) Use hydrogen as carrier gas.
   (d) Adjust the flow rate to 1.12 mL/min.
   (e) Set linear velocity to 39 cm/s.
   (f) Program the oven temperature:
      - Start at 80 °C for 2 min.
      - Ramp up at 10 °C/min to 330 °C.
      - Hold at 330 °C for 6 min.
   (g) Set transfer line temperature to 310 °C.
   (h) Set the ion source temperature to 280 °C.
   (i) Use the scan rate of 10,000 u/s.
   (j) Record mass spectra in the range m/z 85—500.
4. Sample Injection
   (a) Inject 1 μL of each derivatized sample into GC-MS, including the procedural blanks.
   (b) Randomize the injection order to minimize systematic bias.

4 Notes

1. Prepare the mixed solvent stock before analysis for an easier 9-multiplication volume (following the 5:2:2 ratio). For example, prepare 36 mL for 30 samples, 72 mL for 70 samples, and so on.
2. Mixed solvents kept for over a month should be replaced with new ones.
3. Ribitol, which is used as an internal standard, needs to be optimized based on the sample. The choice of ribitol needs to be based on the group similarity with the samples, but it is not present in the sample. For example, ribitol is a group of sugar alcohols commonly found in fruit samples. However, Ribitol is not found naturally in the sample. Therefore, it was chosen as the internal standard for fruit samples.
4. For homogeneity, mixed solvents need to be prepared in the order of hydrophilic to hydrophobic (Ultrapure water > Methanol > Ribitol > Chloroform).

5. Dried fruit samples that have recently arrived should be stored in a refrigerator at −30 °C to preserve the metabolites until analysis.
6. An alternative for researchers who do not own a multi-beads shocker is a mortar and pestle with liquid nitrogen.
7. The utilization of multi-bead shockers was more prominent than that of food processors or juicers due to their ability to crush the sample under cold conditions, thus preventing the deterioration of metabolites caused by heat during the homogenization process.
8. Fruit sample weight for GC/MS analysis can be optimized between 5 mg and 10 mg. For sugar analysis, the smaller the amount, the better for keeping the GC/MS working in good condition. If we removed the sugar from GC/MS analysis (whether through SPE or filament-off analysis during sugar peak), 10 mg would be preferable to enhance other metabolite peak detection.
9. Fruit samples are high in sugar content; therefore, several researchers use solid-phase extraction (SPE) to remove the sugar content during GC/MS analysis.
10. Put the samples in a balance area using the Thermo-shaker and centrifuge to ensure the same forces are applied to the sample.
11. During the evaporator centrifuge steps, ensure the ethanol in the machine is filled until it submerges in the coil. If not, the sample will not fully evaporate, hindering the derivatization steps.
12. After the evaporation, the extract can be kept at 4 °C for 24 h until the derivatization procedure.
13. It is better to use the freshly prepared methoxyamine in a pyridine solution.
14. During the derivation process, the new ampule of MSTFA will give better results than the old/opened one.
15. The GC/MS instrument is sensitive; therefore, it will be better to centrifuge after the silylation step to ensure that the sample moved to the GC/MS vial is clean of debris.

References

1. Lobo FA, Nascimento MA, Domingues JR, Falcão DQ, Hernanz D, Heredia FJ, de Lima Araujo KG (2017) Foam mat drying of Tommy Atkins mango: effects of air temperature and concentrations of soy lecithin and carboxymethylcellulose on phenolic composition, mangiferin, and antioxidant capacity. Food Chem 221:258–266. https://doi.org/10.1016/j.foodchem.2016.10.080
2. Ullah MA, Khanal A, Joyce P, White N, Macnish A, Joyce D (2024) Internal disorders of mango fruit and their management—physiology, biochemistry, and role of mineral nutrients. Plants 13(18). https://doi.org/10.3390/plants13182596
3. Ratti C (2001) Hot air and freeze-drying of high-value foods: a review. J Food Eng 49(4):

311–319. https://doi.org/10.1016/S0260-8774(00)00228-4

4. Hasan MU, Malik AU, Ali S, Imtiaz A, Munir A, Amjad W, Anwar R (2019) Modern drying techniques in fruits and vegetables to overcome postharvest losses: a review. J Food Process Preserv 43(12):e14280. https://doi.org/10.1111/jfpp.14280
5. Omolola AO, Jideani AIO, Kapila PF (2017) Quality properties of fruits as affected by drying operation. Crit Rev Food Sci Nutr 57(1):95–108. https://doi.org/10.1080/10408398.2013.859563
6. Shah AS, Bhat SV, Muzaffar K, Ibrahim SA, Dar BN (2022) Processing technology, chemical composition, microbial quality and health benefits of dried fruits. Curr Res Nutr Food Sci 10(1):71–84. https://doi.org/10.12944/CRNFSJ.10.1.06
7. Yulni T, Agusta W, Jayanegara A, Alfa MN, Hartono LK, Putri Mariastuty TE, Hermansyah HD, Astuti F, Anggraeni D, Lintang MMJ (2024) Unveiling the influence of osmotic pretreatment on dried fruit characteristics: a meta-analysis approach. Prev Nutr Food Sci 29(2):178–189. https://doi.org/10.3746/pnf.2024.29.2.178
8. Sato M, Ikram MMM, Pranamuda H, Agusta W, Putri SP, Fukusaki E (2021) Characterization of five Indonesian mangoes using gas chromatography–mass spectrometry-based metabolic profiling and sensory evaluation. J Biosci Bioeng 132(6):613–620. https://doi.org/10.1016/j.jbiosc.2021.09.006
9. Dettmer K, Aronov PA, Hammock BD (2007) Mass spectrometry-based metabolomics. Mass Spectrom Rev 26(1):51–78. https://doi.org/10.1002/mas.20108
10. Bligh E, Dyer W (1959) A rapid method of total lipid extraction and purification. Can J Biochem Physiol 37:911–917. https://doi.org/10.1139/o59-099
11. Mahieu NG, Patti GJ (2017) Systems-level annotation of a metabolomics data set reduces 25,000 features to fewer than 1,000 unique metabolites. Anal Chem 89(19):10397–10406. https://doi.org/10.1021/acs.analchem.7b02380
12. Kim HK, Verpoorte R (2010) Sample preparation for plant metabolomics. Phytochem Anal 21(1):4–13. https://doi.org/10.1002/pca.1188
13. Kind T, Wohlgemuth G, Lee DY, Lu Y, Palazoglu M, Shahbaz S, Fiehn O (2009) FiehnLib: mass spectral and retention index libraries for metabolomics based on quadrupole and time-of-flight gas chromatography/mass spectrometry. Anal Chem 81(24):10038–10048. https://doi.org/10.1021/ac9019522
14. Villas-Bôas SG, Smart KF, Sivakumaran S, Lane GA (2011) Alkylation or silylation for analysis of amino and non-amino organic acids by GC-MS? Metabolites 1(1):3–20. https://doi.org/10.3390/metabo1010003
15. Miyagawa H, Bamba T (2019) Comparison of sequential derivatization with concurrent methods for GC/MS-based metabolomics. J Biosci Bioeng 127(2):160–168. https://doi.org/10.1016/j.jbiosc.2018.07.015
16. Tandel J, Tandel Y, Kapadia C, Singh S, Gandhi K, Datta R, Singh S, Yirgu A (2023) Nontargeted metabolite profiling of the most prominent Indian mango (*Mangifera indica* L.) cultivars using different extraction methods. ACS Omega 8(43):40184–40205. https://doi.org/10.1021/acsomega.3c03670
17. Ganneru S, Shaik H, Peddi K, Mudiam MKR (2019) Evaluating the metabolic perturbations in *Mangifera indica* (mango) ripened with various ripening agents/practices through gas chromatography–mass spectrometry-based metabolomics. J Sep Sci 42(19):3086–3094. https://doi.org/10.1002/jssc.201900291

Chapter 12

GC-MS-Based Metabolite Profiling Method for Dried Vegetables

Waqif Agusta

Abstract

This chapter presents a comprehensive protocol for the extraction, derivatization, and GC-MS analysis of metabolites in fresh and fried vegetables, focusing on sweet potato (*Ipomoea batatas*) and broccoli (*Brassica oleracea var. italica*). These vegetables are nutritionally important and increasingly incorporated into processed products through vacuum and deep-frying techniques. However, these thermal processes significantly alter their metabolite profiles, particularly sugars, amino acids, and organic acids. The protocol includes optimized sample preparation, Bligh-Dyer extraction, and dual-step derivatization (methoximation followed by silylation using MSTFA) to enhance volatility and detectability in GC-MS. It also accounts for matrix-specific challenges, including lipid interference in fried samples and moisture variability. Applications of this method include nutritional profiling, detection of process-induced transformations, and quality control in vegetable-based snacks. This workflow provides a practical, adaptable guide for researchers conducting high-resolution metabolite analysis in plant-derived food matrices.

Key words GC-MS, Metabolomics, Vegetable, Sweet potato, Broccoli, Extraction, Derivatization

1 Introduction

Vegetables are essential for a well-balanced diet, providing rich vitamins, minerals, dietary fiber, and phytochemicals that support overall health [1, 2]. Sweet potato (*Ipomoea batatas*) and broccoli (*Brassica oleracea var. italica*) are notable for their unique nutritional compositions and diverse culinary uses. Sweet potato is a nutritionally rich root vegetable containing the source of complex carbohydrates, dietary fiber, and various vitamins and minerals [3]. Broccoli is a highly nutritious cruciferous vegetable renowned for its health-promoting properties. It is rich in vitamins (A, C, K), minerals, fiber, and bioactive compounds such as glucosinolates and phenolics [4]. Both vegetables also contain a wide spectrum of metabolites—such as sugars, organic acids, and amino acids—that play essential roles in plant physiology, taste, and human nutrition.

Sastia Prama Putri (ed.), *Gas Chromatography-Mass Spectrometry: Based Metabolomics in Food Science*, Methods and Protocols in Food Science, https://doi.org/10.1007/978-1-0716-5108-7_12,

The increasing demand for vegetable-based convenience products has driven innovation in food preservation and processing technologies. Among these, vacuum and deep-frying have gained popularity as methods for producing shelf-stable vegetable snacks. Vacuum frying, which involves cooking at reduced pressure and lower temperatures, is especially valued for minimizing oil absorption, preserving color and nutrients, and reducing the formation of harmful thermal degradation products [5, 6]. Deep-fat frying is a widely used cooking method, which involves higher temperatures and atmospheric pressure, often leading to more intense thermal reactions including oxidation, hydrolysis, isomerization, and polymerization, which affect the final product's quality, flavor, and nutritional composition [7].

The metabolite fraction of vegetables includes reducing sugars (e.g., glucose, fructose), disaccharides (e.g., sucrose), organic acids (e.g., citric, malic, fumaric), and amino acids (e.g., asparagine, glutamic acid, alanine). These compounds are important as nutrients and flavor contributors and also as precursors to chemical transformations during frying. For instance, the existence of reducing sugars and amino acids during high-temperature cooking may not only cause the Maillard reaction, produce flavor compounds and browning pigments but also potentially form undesirable by-products such as acrylamide [8, 9]. Moreover, organic acids influence the acidity and taste of the vegetables [10], and their thermal degradation or volatilization can alter the perceived quality of fried products.

Comprehensive analytical profiling is essential to understanding the impact of frying on these valuable metabolites. However, accurate quantification of compounds in complex food matrices such as fried vegetables presents significant analytical challenges. These compounds are typically nonvolatile and thermally labile, making them unsuitable for direct analysis by gas-phase techniques such as Gas Chromatography–Mass Spectrometry (GC-MS). Therefore, targeted analysis of metabolites using GC-MS requires chemical derivatization to convert them into volatile, thermally stable forms [11].

Among derivatization techniques, silylation, especially with reagents like N-methyl-N-(trimethylsilyl)trifluoroacetamide (MSTFA), is the most commonly used approach for metabolite analysis [12–15]. Silylation replaces active hydrogen atoms in hydroxyl, carboxyl, and amino functional groups with trimethylsilyl (TMS) groups, resulting in derivatives with improved volatility and GC compatibility. This reaction is typically carried out under mild conditions, though it is highly sensitive to moisture, which can interfere with the efficiency of derivatization. Complete drying of the extract is therefore essential before adding derivatizing reagents.

The extraction step preceding derivatization is equally critical. Extraction of metabolites from fresh and fried vegetable matrices must be tailored to overcome the differences in moisture content, lipid content, and matrix density. In fresh vegetables, water content is high, and cell walls are intact, requiring physical disruption (e.g., grinding or homogenization) to release intracellular metabolites. In fried samples, lipid residues and reduced water content necessitate additional steps such as defatting with nonpolar solvents or using biphasic systems to separate polar analytes from interfering lipids [16]. The choice of solvent for polar metabolite extraction often includes methanol, ethanol, or water, or combinations thereof, adjusted for polarity and compatibility with downstream derivatization [17, 18]. Vacuum-fried and deep-fried products also present unique challenges in analysis due to the complex physicochemical reactions occurring during the frying process [19], which impact the thermal degradation of sensitive molecules. To mitigate these issues, optimized extraction protocols must consider factors such as extraction time, temperature, and solvent-to-sample ratio and incorporate internal standards for quantification and quality control [20]. Inadequate extraction or incomplete derivatization can compromise GC-MS measurements' sensitivity, resolution, and reproducibility, especially when working with low-abundance metabolites or complex matrices.

Importantly, metabolite profiling is not only useful for understanding nutritional content but also for evaluating process-related transformations, authenticity of raw materials, and quality control in processed vegetable products [21–24]. For example, a decline in asparagine levels during frying may indicate Maillard consumption, while the appearance of degradation products such as furfural may signal excessive thermal stress [21]. Likewise, shifts in sugar composition (e.g., reduction in sucrose and accumulation of glucose) can reflect enzymatic or thermal breakdown, which in turn affects sweetness and browning behavior [22].

Despite the potential of GC-MS for such applications, there is still a lack of standardized, validated workflows for the targeted profiling of metabolites in processed vegetables, particularly those subjected to vacuum and deep frying. Most studies focus on lipid oxidation, vitamin retention, or surface oil content. At the same time, the primary biochemical landscape remains underexplored, especially in terms of absolute quantification and mechanistic understanding of frying effects. Given the increasing health focus on minimally processed, plant-based foods, there is a clear need to expand analytical capabilities to cover nutritional markers and processing indicators in vegetable-based snacks and products.

This chapter aims to fill this gap by providing a detailed and practical overview of extraction and derivatization methods for GC-MS analysis of compounds in fresh and fried vegetables, specifically on sweet potato and broccoli. Emphasis is placed on tailoring protocols to address the compositional differences between raw and fried matrices, optimize extraction conditions for metabolite recovery, and ensure efficient GC-MS detection derivatization. In doing so, the chapter offers methodological guidance and scientific rationale for researchers and practitioners interested in high-quality metabolite analysis in plant-based food systems.

2 Materials and Instruments

2.1 Materials

2.1.1 Sample Preparation

1. Liquid nitrogen.

2.1.2 Extraction

1. Microtube 2 mL and 1.5 mL.
2. Holed-microtube caps.
3. Mixed solvent (methanol: chloroform: ultrapure water) (5:2:2, ratio v/v/v) (*see* **Notes 1 and 2**).
4. Ribitol dissolved in ultrapure water (100 μg/mL) (*see* **Note 3 and 4**).

2.1.3 Derivatization

1. Methoxyamine hydrochloride in pyridine (20 mg/mL).
2. N-methyl-N-(trimethylsilyl)trifluoroacetamide (MSTFA) ampule.

2.2 Instruments

1. Multi-beads shocker with metal beads.
2. Weighing balance.
3. Micropipette.
4. Vortex machine.
5. Thermomixer incubator with shaker mechanism.
6. Refrigerated centrifuge.
7. Centrifugal evaporator.
8. Freeze dryer.

3 Methods

Vegetable samples, such as sweet potato and broccoli, have matrix properties that differ from fruits, including lower sugar content, higher moisture, and fiber levels. Therefore, the extraction protocol was optimized to accommodate these characteristics without needing separate analytical workflows.

3.1 Sample Preparation

1. Fresh and processed vegetable samples, such as sweet potato and broccoli, were finely chopped and cryogenically frozen using liquid nitrogen and freeze-dried overnight (*see* **Note 5**)
2. Samples were then pulverized into a fine powder using a multi-bead shocker (*see* **Notes 6 and** 7)

3.2 Extraction

1. Weigh 5 mg of powdered sample into a 2 mL microtube (*see* **Note 8**).
2. Add 1 mL of mixed solvent containing ribitol (100 μg/mL) as internal standard using a micropipette, and then vortex the sample to ensure homogeneity (*see* **Note 9**)
3. Prepare procedure blanks with solvent only.
4. Incubate in the Thermomixer incubator under the conditions of 37 °C, 1200 rpm for 30 min. (*see* **Note 10**)
5. Vortex and centrifuge the samples for 3 min at 4 °C, 10,000 rpm for separating the solid and liquid phase (*see* **Note 10**)
6. Transfer 600 μL of supernatant to a new 1.5 mL microtube.
7. Add 300 μL of ultrapure water, vortex, and centrifuge the sample again for 3 min at 4 °C, 10,000 rpm to homogenize, and separate the hydrophilic and hydrophobic phases (*see* **Note 10**).
8. Separate the supernatants into two parts. Transfer 100 μL of supernatant to a new 1.5 mL microtube, collect another 100 μL of supernatant, and gather it in the QC pool (2 mL microtube).
9. To make QC samples, transfer 100 μL of mixed supernatant from all samples (QC pool) into a new 1.5 mL microtube.
10. Close all samples with holed caps and put the microtube into the vacuum evaporator to evaporate the solvent for 1 h at room temperature until no water is left (*see* **Notes 10** and **11**).
11. Store extracts at 4 °C until derivatization (*see* **Note 12**).

3.3 Derivatization

1. Methoxyamine hydrochloride 20 mg/mL in pyridine was added to the lyophilized sample for 100 μL and closed again with standard caps (*see* **Note 13**).
2. Incubate the first-derivatized sample in Thermomixer for 90 min at 30 °C, 1200 rpm.
3. Add 50 μL of MSTFA to the sample (*see* **Note 14**)
4. Incubate the second-derivatized sample in Thermomixer for 30 min at 37 °C, 1200 rpm.
5. Centrifuge the sample for 3 min at 25 °C, 10,000 rpm for separating the debris from the supernatant (*see* **Note 15**)
6. Transfer 100 μL of derivatized sample to the GC/MS vial.

3.4 GC-MS Analysis

3.4.1 Instrument Preparation

1. Use either GC-MS-QP2010 Ultra or GC-MS-TQ8030 (Shimadzu, Kyoto, Japan) equipped with an autosampler AOC-20i/s (Shimadzu, Kyoto, Japan).
2. Install a guard column: InertCap 5MS/NP ProGuard, 35 m × 0.18 mm I.D., 0.18 μm df; GL Sciences.

3.4.2 Tuning and Calibration

1. Tune and calibrate the mass spectrometer before sample analysis.
2. Inject a standard alkane mixture (C_9-C_{40}) to enable calculation of linear retention indices (RI).

3.4.3 GC-MS Settings

1. Set the injection mode to split (25:1, v/v).
2. Set the injection temperature to 270 °C.
3. Use hydrogen as a carrier gas.
4. Adjust the flow rate to 1.48 mL/min.
5. Set linear velocity to 60 cm/s.
6. Program the oven temperature:
   - Start at 80 °C for 2 min.
   - Ramp up at 10 °C/min to 330 °C.
   - Hold at 330 °C for 6 min.
7. Set transfer line temperature to 290 °C.
8. Set the ion source temperature to 280 °C.
9. Use the scan rate of 10000 u/s.
10. Record mass spectra in the range m/z 85–500.

3.4.4 Sample Injection

1. Inject one μL of each derivatized sample into the GC-MS, including the procedural blanks.
2. Randomize the injection order to minimize systematic bias.

4 Notes

1. Prepare the mixed solvent stock prior to analysis for an easier nine-multiplication volume (following the 5:2:2 ratio). For example, prepare 36 mL for 30 samples, 72 mL for 70 samples, and so on.
2. Mixed solvents that have been kept for more than 1 month should better replaced with a new one.
3. Ribitol, which is used as an internal standard, needs to be optimized based on the sample. The choice of ribitol needs to

be based on the group similarity with the samples, but it is not present in the sample. For example, ribitol is a group of sugar alcohols commonly found in fruit samples. However, ribitol is not found naturally in the sample. Therefore, it was chosen as the internal standard for fruit samples.

4. For homogeneity, mixed solvents need to be prepared in the order of hydrophilic to hydrophobic (ultrapure water > methanol > ribitol > chloroform).
5. Samples that have recently arrived should be stored in a refrigerator at −30 °C to preserve the metabolites until analysis.
6. An alternative for researchers who do not own a multi-bead shocker is a mortar and pestle with liquid nitrogen.
7. Utilization of multi-bead shockers was more prominent compared to food processors or juicers due to their ability to crush the sample under cold conditions, thus preventing deterioration of metabolites caused by heat during the homogenization process.
8. GC/MS analysis sample weight can be optimized between 5 mg and 10 mg. For sugar analysis, the smaller the amount, the better for keeping the GC/MS working in good condition. If we removed the sugar from GC/MS analysis (whether through SPE or filament-off analysis during sugar peak), 10 mg would be preferable to enhance other metabolite peak detection.
9. Samples that are high in sugar content; therefore, several researchers use solid-phase extraction (SPE) to remove the sugar content during GC/MS analysis.
10. Put the samples in a balance area using the thermomixer and centrifuge to ensure the same forces are applied to the sample.
11. During the evaporator centrifuge steps, ensure the ethanol in the machine is filled until it submerges in the coil. If not, the sample will not be fully evaporated, hindering the derivatization steps.
12. After the evaporation, the extract can be kept at 4 °C maximum of 24 h until the derivatization procedure.
13. It is better to use the freshly prepared methoxyamine in a pyridine solution.
14. During the derivation process, the new ampule of MSTFA will give better results than the old/opened one.
15. The GC/MS instrument is sensitive; therefore, it will be better to centrifuge after the silylation step to ensure that the sample that will be moved to the GC/MS vial is clean of debris.

References

1. Dias JS, Dias JS (2012) Nutritional quality and health benefits of vegetables: a review. Food Nutr Sci 3(10):1354–1374. https://doi.org/10.4236/fns.2012.310179
2. Moreira RG (2014) Vacuum frying versus conventional frying - an overview. Eur J Lipid Sci Technol 116(6):723–734. https://doi.org/10.1002/ejlt.201300272
3. Laveriano-Santos EP, López-Yerena A, Jaime-Rodríguez C, González-Coria J, Lamuela-Raventós RM, Vallverdú-Queralt A, Romanyà J, Pérez M (2022) Sweet potato is not simply an abundant food crop: a comprehensive review of its phytochemical constituents, biological activities, and the effects of processing. Antioxidants 11(9):1648. https://doi.org/10.3390/antiox11091648
4. Garcia J, Syed RU, Moni SS, Khaled Bin Break M, Khojali WMA, Jafar M, Alshammari MD, Abdelsalam K, Taymour S, Saad K, Alreshidi M, Mohamed M, Taha E, Mohan S (2023) Broccoli: a multi-faceted vegetable for health: an in-depth review of its nutritional attributes, antimicrobial abilities, and anti-inflammatory properties. Antibiotics 12(7):1157. https://doi.org/10.3390/antibiotics12071157
5. da Silva PF, Moreira RG (2008) Vacuum frying of high-quality fruit and vegetable-based snacks. LWT 41(10):1758–1767. https://doi.org/10.1016/j.lwt.2008.01.016
6. Dueik V, Bouchon P (2011) Vacuum frying as a route to produce novel snacks with desired quality attributes according to new health trends. J Food Sci 76(2):E188–E195. https://doi.org/10.1111/j.1750-3841.2010.01976.x
7. Zhang Q, Saleh ASM, Chen J, Shen Q (2012) Chemical alterations taken place during deep-fat frying based on certain reaction products: a review. Chem Phys Lipids 165(6):662–681. https://doi.org/10.1016/j.chemphyslip.2012.07.002
8. Mottram D, Wedzicha B, Dodson A (2002) Acrylamide is formed in the Maillard reaction. Nature 419:448–449. https://doi.org/10.1038/419448a
9. Stadler R, Blank I, Varga N et al (2002) Acrylamide from Maillard reaction products. Nature 419:449–450. https://doi.org/10.1038/419449a
10. Shi Q, Han G, Liu Y, Jiang J, Jia Y, Li X (2022) Nutrient composition and quality traits of dried jujube fruits in seven producing areas based on metabolomics analysis. Food Chem 385:132627. https://doi.org/10.1016/j.foodchem.2022.132627
11. Dettmer K, Aronov PA, Hammock BD (2007) Mass spectrometry-based metabolomics. Mass Spectrom Rev 26(1):51–78. https://doi.org/10.1002/mas.20108
12. Afifah EN, Sari IA, Susilo AW, Malik A, Fukusaki E, Putri SP (2024) Characterization of fine-flavor cocoa in parent-hybrid combinations using metabolomics approach. Food Chemistry: X 24:101832. https://doi.org/10.1016/j.fochx.2024.101832
13. Ati AG, Sari IA, Firmanto H, Malik A, Fukusaki E, Putri SP (2025) Effect of pre-drying on flavor modulation in Indonesian cocoa beans: a metabolomics study of key flavor compounds and sensory profiles. Food Biosci 65:106056. https://doi.org/10.1016/j.fbio.2025.106056
14. Ikram MMM, Ridwani S, Putri SP, Fukusaki E (2020) GC-MS based metabolite profiling to monitor ripening-specific metabolites in pineapple (*Ananas comosus*). Meta 10(4):134. https://doi.org/10.3390/metabo10040134
15. Sato M, Ikram MMM, Pranamuda H, Agusta W, Putri SP, Fukusaki E (2021) Characterization of five Indonesian mangoes using gas chromatography–mass spectrometry-based metabolic profiling and sensory evaluation. J Biosci Bioeng 132(6):613–620. https://doi.org/10.1016/j.jbiosc.2021.09.006
16. Kim HK, Verpoorte R (2010) Sample preparation for plant metabolomics. Phytochem Anal 21(1):4–13. https://doi.org/10.1002/pca.1188
17. Masson P, Alves AC, Ebbels TMD, Nicholson JK, Want EJ (2010) Optimization and evaluation of metabolite extraction protocols for untargeted metabolic profiling of liver samples by UPLC-MS. Anal Chem 82(18):7779–7786. https://doi.org/10.1021/ac101722e
18. Yanes O, Tautenhahn R, Patti GJ, Siuzdak G (2011) Expanding coverage of the metabolome for global metabolite profiling. Anal Chem 83(6):2152–2161. https://doi.org/10.1021/ac102981k
19. Zhang Q, Qin W, Li M, Shen Q, Saleh ASM (2015) Application of chromatographic techniques in the detection and identification of constituents formed during food frying: a review. Compr Rev Food Sci Food Saf 14(5):601–633. https://doi.org/10.1111/1541-4337.12147

20. Mahieu NG, Patti GJ (2017) Systems-level annotation of a metabolomics data set reduces 25,000 features to fewer than 1,000 unique metabolites. Anal Chem 89(19): 10397–10406. https://doi.org/10.1021/acs.analchem.7b02380
21. Huang Y, Li H, Yu X, Zhang F, Wang A, Wan X, Zhuang P, Jiao J, Zhang Y (2025) Foodomics-based chemical profiling and signature compound characterization of air-fried potato chips: the effect of thermal processing and fish oil treatment on advancing safety, nutrition, and flavor attributes. Food Biosci 66:106211. https://doi.org/10.1016/j.fbio.2025.106211
22. Jiang X, Zhang R, Yao Y, Yang Y, Wang B, Wang Z (2023) Effect of cooking methods on metabolites of deep purple-fleshed sweetpotato. Food Chem 429:136931. https://doi.org/10.1016/j.foodchem.2023.136931
23. Wan X, Wang X, Zhu T, Shi G, Xiao Q (2025) Comparative metabolomics analysis of purple sweet potato flesh across different storage periods. Postharvest Biol Technol 229:113730. https://doi.org/10.1016/j.postharvbio.2025.113730
24. Zhao Y, Zhang Y, Yang H, Xu Z, Li Z, Zhang Z, Zhang W, Deng J (2024) A comparative metabolomics analysis of phytochemicals and antioxidant activity between broccoli floret and by-products (leaves and stalks). Food Chem 443:138517. https://doi.org/10.1016/j.foodchem.2024.138517

Chapter 13

GC-MS-Based Volatile Compounds Profiling Method for Jasmine Rice Seedling Leaves

Mongkon Sirijan and Supaart Sirikantaramas

Abstract

This chapter describes the methodologies employed to identify volatile metabolites in jasmine rice seedling leaf tea processed via freeze-drying or sun-drying, using headspace gas chromatography–mass spectrometry. The profiling focuses on key classes of volatile compounds, including aldehydes, hydrocarbons, alcohols, ketones, carboxylic acids, and other miscellaneous constituents. Among these, 2-methylbutanal, dodecane, 2-ethylfuran, and methylpyrazine were identified as major contributors to the tea's characteristic aroma, potentially enhancing its sensory appeal. The chapter further details protocols for metabolite extraction, compound identification, data analysis, quality control, and experimental replication. These analytical approaches may serve as a model for metabolite profiling in other herbal teas, with broader applications in functional beverage development and plant-based product innovation.

Key words GC-MS, Herbal tea, Headspace, *Oryza sativa* L., Seedling

1 Introduction

Tea derived from the leaves of *Camellia sinensis* represents one of the most widely consumed beverages worldwide, renowned for its high antioxidant content and numerous associated health benefits [1]. In addition to traditional tea, various plant parts, such as leaves, flowers, seeds, fruits, and roots, can be processed into herbal teas, which offer distinctive flavors and are rich in bioactive compounds.

Recently, rice (*Oryza sativa* L.) leaf tea has emerged as a novel herbal infusion, garnering attention for its potential health-promoting properties and increasing commercial appeal. As a staple crop throughout Asia and the Pacific [2], rice holds significant global importance. Within the diverse varieties of rice, Thai jasmine rice is particularly esteemed for its characteristic fragrance, soft texture, and notable antioxidant activity. These attributes position it as a promising candidate for the development of functional tea products.

Sastia Prama Putri (ed.), *Gas Chromatography-Mass Spectrometry: Based Metabolomics in Food Science*, Methods and Protocols in Food Science, https://doi.org/10.1007/978-1-0716-5108-7_13,

Building on this foundation, our previous research explored the metabolite composition of jasmine rice seedling leaves using a dual-platform metabolomics approach combining liquid chromatography–mass spectrometry (LC–MS) and gas chromatography–MS (GC–MS). Specifically, we investigated the effects of two common drying methods, i.e., freeze-drying and sun-drying, on the chemical profiles of the leaves [3]. Headspace GC–MS (HS GC–MS) analysis revealed that sun-dried leaves contained a higher relative abundance of key volatile compounds, including 2-methylbutanal, dodecane, 2-ethylfuran, and methylpyrazine. These compounds are known contributors to aroma enhancement and may improve the sensory appeal of rice leaf tea. Notably, several volatiles identified in jasmine rice seedling leaves, such as 2-methylbutanal, propanal, and hexanal, are also found in dried green tea leaves, as reported in prior GC–MS analyses [4], suggesting shared flavor-active constituents that influence consumer preferences. Supporting this, Kim et al. [5] demonstrated that higher levels of 2-methylbutanal in green tea positively influenced consumer acceptance, based on solid-phase microextraction–GC/MS analysis [5]. Collectively, these findings highlight the utility of GC–MS as a robust analytical tool for the identification and quantification of volatile compounds in rice seedling leaves.

This chapter presents a comprehensive GC–MS-based analysis of the metabolite profile of jasmine rice seedling leaf tea, with detailed descriptions of the methodologies employed for sample preparation, compound extraction, data analysis, quality control, and experimental replication. The insights derived from this study advance our understanding of the volatile composition of rice leaf teas and provide a scientific basis for the development of biofortified, health-promoting, plant-based functional beverages. Furthermore, the analytical framework established here may serve as a model for investigating the volatile profiles of other herbal teas using comparable methodologies.

2 Materials

2.1 Plant Sample

Freeze-dried or sun-dried jasmine rice seedling leaves were used as the plant sample for analysis.

2.2 Reagents and Chemicals

Thiophene was used as an internal standard for volatile compound analysis. A standard mixture of *n*-alkane (C_7—C_{30}) was employed for retention index calculations.

2.3 Equipment

The following laboratory equipment and materials were used throughout the experiment: pipettes with disposable tips, 10-mL HS vials, 20-mL aluminum caps, polytetrafluoroethylene/silicone septa, a crimper, an analytical balance with an accuracy of ±0.1 mg, a GC–MS system equipped with an HP-INNOWax capillary GC column (part no. 19091 N-133, 30-m length, 0.25-mm inner diameter, 0.25-μm film thickness), a lyophilizer or freeze-dryer, and a Mixer mill/grinding machine.

3 Methods

3.1 Plant Materials and Sample Preparation

1. Jasmine rice seeds (cv. Hom Dang) were cultivated at 25 ± 2 °C and 70% relative humidity (RH) and harvested 7 days after germination (Fig. 1a). Following harvest, the roots were removed and discarded, and the seedling leaves were thoroughly washed with distilled water and dried using tissue paper to remove surface moisture.
2. The 7-day-old jasmine rice seedling leaf tissues were either placed in a 50-mL tube and immediately immersed in liquid nitrogen to halt metabolic activity and enzyme activity or directly processed for volatile compound analysis (*see* **Note 1**).
3. The rice leaves were then subjected to one of the two drying methods: freeze-drying and sun-drying.
   - For freeze-drying, the leaves were placed in a freeze-dryer and lyophilized overnight in a vacuum flask at 9.7 mTorr and −56.5 °C (Fig. 1b).
   - For sun-drying, the leaves were placed on aluminum trays and exposed to natural sunlight (25 °C–35 °C) for approximately 3 days until the moisture content dropped below 10% on a wet basis (Fig. 1c).
4. The dried jasmine rice seedling leaves were ground into a fine powder using a Mixer mill equipped with a stainless-steel chamber and a 15-mm grinding ball (*see* **Note 2**).

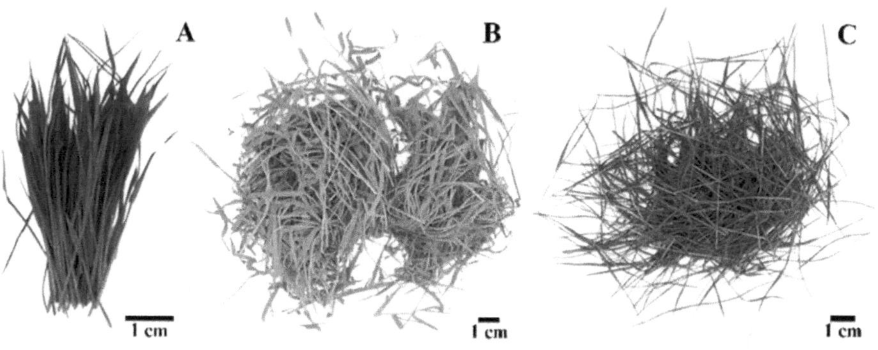

Fig. 1 Jasmine rice seedling leaves (**a**) harvested 7 days after germination and subjected to (**b**) freeze-drying and (**c**) sun-drying

3.2 HS-GC-MS Analysis

1. The volatile compound analysis was performed using static HS GC–MS, following the procedure depicted in Fig. 2. Briefly, 0.5 g of dried jasmine rice seedling leaf powder was weighed and transferred to a 10-mL HS vial. After adding 2 μL of a 1000 μg/mL thiophene stock solution as the internal standard, the vial was immediately sealed and placed in the HS oven for analysis (*see* **Note 3**).
2. For the HS GC–MS analysis, an Agilent 7890B GC system coupled with an Agilent 7000C triple quad mass spectrometer and an Agilent 7697A static HS autosampler was used, employing electron impact ionization (EI) for MS. Analyte separation was achieved using an HP-INNOWax capillary column (part no. 19091N-133, 30-m length, 0.25-mm inner diameter, and 0.25-μm film thickness) with a specific oven temperature program. Please note that equivalent instrumentation with comparable specifications may also be suitable. Detailed HS and GC conditions are provided in Tables 1 and 2.

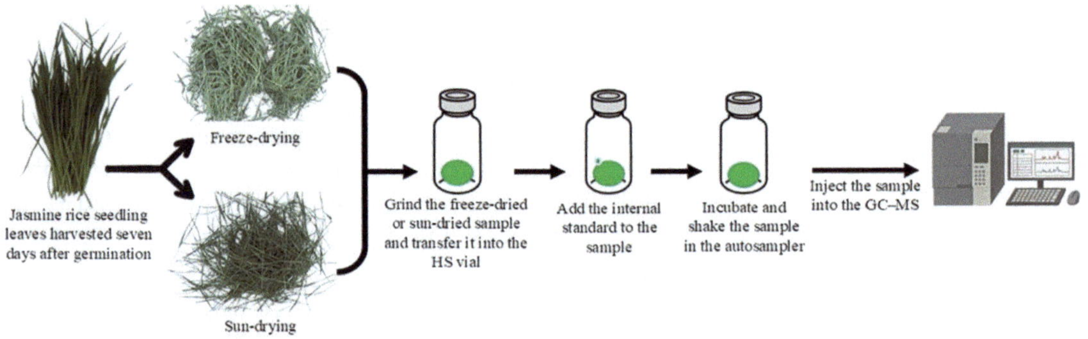

Fig. 2 Flowchart illustrating the HS GC–MS analysis procedure for jasmine rice seedling leaves

Table 1
Headspace parameters

Parameter		Condition
Oven temperature		120 °C
Oven	Loop temperature	150 °C
	Transfer line temperature	180 °C
	Equilibration time	9 min
Time	Injection time	1 min
	Loop size/injection volume	1 mL
	GC cycle time	72 min

Table 2
GC parameters for volatile analysis

Parameter		Condition
Injection	Injection port temperature	220 °C
	Inlet mode	Split ratio (20:1)
Column	HP-INNOWax column	
	Helium carrier flow rate	1 mL/min (constant flow)
Oven (initial)	Initial temperature	40 °C
	Hold time	4 min
Oven (Ramp1)	Ramp temperature	250 °C
	Ramp rate	5 °C/min
	Hold time	4 min
Post-run	Temperature	260 °C
	Hold time	5 min
MS detector	Solvent delay	0 min
	MS scan range	29–400 m/z
	Transfer line temperature	250 °C
	EI source temperature	230 °C
	EI voltage	70 eV

3.3 Data Analysis

The following steps were performed to analyze the data:

1. Retention indices were calculated using the *n*-alkane standard mixture (C_7–C_{30}) to ensure accuracy and consistency.
2. The MS spectra were analyzed to annotate compounds by comparing the observed mass-to-charge (*m/z*) ratios with reference spectra obtained from standards or reference databases such as the National Institute of Standards and Technology or Wiley MS library. Only compounds with a spectral similarity of >80% were considered reliably identified.
3. The annotated volatile metabolites were quantified by normalizing their peak areas to that of the internal standard (thiophene). The results were expressed as relative metabolite levels.
4. Statistical analysis was performed using SPSS software. For multivariate analyses, the normalized peak intensities of metabolites from all analytical platforms were processed using MetaboAnalyst 6.0 [6].

3.4 Quality Control and Replication

1. Blank controls (empty HS vials) were included in each analytical batch to monitor background effects. Internal standards were incorporated into each sample batch to ensure analytical accuracy and reproducibility across runs.
2. To ensure accurate GC–MS results, an appropriate sample distribution within the vial is crucial. Each sample was

transferred to a GC–MS vial of appropriate volume to allow effective access by the autosampler injection needle. In addition, the injection order of samples was randomized to minimize potential biases associated with systematic variations between samples.

3. At least three biological and three technical replicates were performed for each plant sample.

4 Notes

1. Please note that direct contact between the sample and liquid nitrogen should be avoided. The tube lid should be securely fastened to prevent leakage; however, liquid nitrogen should not be trapped inside the Falcon tube to prevent pressure buildup and potential explosion.
2. Dried jasmine rice seedling leaves were ground into a fine powder using an MM400 Mixer mill (Retsch®, Germany) equipped with a 0.35-mm (45 mesh) sieve at 30 Hz for 1 min.
3. A 1000 μg/mL thiophene solution was prepared in LC-grade water. Subsequently, 0.5 g of dried jasmine rice seedling leaf powder was placed into an HS vial, followed by the addition of 2 μL of the thiophene solution. The vial was immediately sealed to prevent the loss of volatiles.

Acknowledgments

Mongkon Sirijan was financially supported by the University Personnel Development Fund 2001 (No. 11/2567), Naresuan University.

Competing Interests The authors have no conflicts of interest to declare that are relevant to the content of this chapter.

References

1. Kome D, Horzic D, Belscak A, Ganic KK, Vulic I (2010) Green tea preparation and its influence on the content of bioactive compounds. Food Res Int 43:167–176. https://doi.org/10.1016/j.foodres.2009.09.022
2. Verma DK, Srivastav PP (2020) A paradigm of volatile aroma compounds in rice and their product with extraction and identification methods: a comprehensive review. Food Res Int 130: 108924. https://doi.org/10.1016/j.foodres.2019.108924
3. Sirijan M, Hutasingh N, Khaksar G, Pewlong P, Pattarapipatkul N, Tubtimrattana A, Tansrisawad N, Borompichaichartkul C, Sirikantaramas S (2025) Uncovering bioactive flavones and volatiles in rice (*Oryza sativa* L.) leaf tea through dual-platform metabolomics. LWT

Food Sci Technol 218:117499. https://doi.org/10.1016/j.lwt.2025.117499

4. Flaig M, Qi S, Wei G, Yang X, Schieberle P (2020) Characterization of the key odorants in a high-grade Chinese green tea beverage (*Camellia sinensis*; Jingshan cha) by means of the sensomics approach and elucidation of odorant changes in tea leaves caused by the tea manufacturing process. J Agric Food Chem 68: 5168–5179. https://doi.org/10.1021/acs.jafc.0c01300
5. Kim Y, Lee KG, Kim MK (2016) Volatile and non-volatile compounds in green tea affected in harvesting time and their correlation to consumer preference. J Food Sci Technol 53: 3735–3743. https://doi.org/10.1007/s13197-016-2349-y
6. Pang Z, Lu Y, Zhou G, Hui F, Xu L, Viau C, Spigelman AF, MacDonald PE, Wishart DS, Li S, Xia J (2024) MetaboAnalyst 6.0: towards a unified platform for metabolomics data processing, analysis and interpretation. Nucleic Acids Res 52:W398–W406. https://doi.org/10.1093/nar/gkae253

Part III

Legumes and Their Derived Products

Chapter 14

GC-MS-Based Metabolite Profiling Method for Legumes and Tempe, an Indonesian Fermented Food

Della Rahmawati

Abstract

Metabolomics, a comprehensive analysis of metabolites in biological sample, has been exhaustively used in food science and nutrition research, including research about legumes. Legumes such as soybean, edamame, and other beans have medicinal benefits that are reflected in their metabolites, such as essential amino acids, fibers, and isoflavones. Not only in the legumes themselves, but the product derivatives also show to have numerous medicinal benefits, such as tempe, that are able to replace meat as a protein source. One of the instruments that has been extensively used for metabolomics of legumes is gas chromatography/mass spectrometer (GC/MS). Two treatments that are able to help legume samples to be detected in GC/MS are extraction and derivatization. The extraction that will be discussed here is based on the Bligh-Dyer method, while the derivatizations are oximation and silylation. Several research have been conducted for analyzing legume-based products using a GC/MS instrument, such as bean analysis, fermentation process in legumes, and different metabolite profiles of legumes. The detailed methodology for analyzing legumes and legume-based products is depicted in this book chapter to let the readers understand the methodology that needs to be conducted to ensure high-quality data for legume GC-MS analysis.

Key words Legume, Metabolomics, Gas chromatography-mass spectrometry, Oximation, Silylation

1 Introduction

Metabolomics, or the global analysis of metabolites in biological material, has been extensively applied to food science and nutrition research. Metabolomics facilitates the detailed analysis of small molecules that define the compositional, quality, or functional properties of food materials. Within these food materials, legumes hold a prominent role of their nutritional as well as commercial value. For example, edamame (immature green soybeans) contains a greater amount of protein, essential amino acids, dietary fiber, isoflavones, and other bioactive compounds, making them a primary target in metabolism profiling in order to maximize their potential in functional food and nutraceutical applications. Soybean is a significant source of many metabolites, so it is an important

Sastia Prama Putri (ed.), *Gas Chromatography-Mass Spectrometry: Based Metabolomics in Food Science*, Methods and Protocols in Food Science, https://doi.org/10.1007/978-1-0716-5108-7_14,

nutritional resource [1]. A key component of their composition is protein, with soybeans compared with other legumes, and the presence of two subcategories, glycinin and β-conglycinin, which lead to their nutritional attributes. During digestion or processing, they can also generate bioactive peptides with potential health benefits, such as antioxidant and antihypertensive activity [2]. Isoflavones, which are natural phytoestrogens, represent another important class of compounds in soybeans. These are mainly present as their glucoside forms, including genistein, daidzein, and glycitein, and are linked to several health-promoting effects, including bone health and can reduce menopausal symptoms [3].

The metabolism of the microbes, such as *Rhizopus oligosporus*, which are responsible for fermenting the soybeans into tempeh, greatly alters the nutritional profile of tempeh. Fermentation breaks down soy proteins and increases digestibility by making proteins into smaller peptides and free amino acids. Not only do they ease the digestion of the proteins, but they also bring about the generation of bioactive peptides, which have many health benefits [4].

Isoflavones are bonded to sugars (in their glucoside form) in raw soybeans but are converted into their aglycone form in fermented products. This process makes them more bioavailable, and the body is then in a much better position to absorb and utilize them. This makes tempeh much healthier, providing benefits such as better gut health through its probiotic and prebiotic content, as well as more diverse cardiovascular benefits from bioactive peptides and antioxidants.

Tempeh is an easily assimilated, nutritionally enhanced product as compared to soybeans. The metabolites processed and generated during fermentation render tempeh a functional food, which can support health on multiple levels: digestive (improved digestive comfort), health (enhanced bioavailability), and disease (potential protective effects against chronic disease). Therefore, advanced analytical techniques are needed to elucidate both the metabolite profiles of legume seed and their derivatives. The gas chromatography-mass spectrometry (GC-MS) platform is gaining new traction as a preferred solution based on high sensitivity, accuracy, and ability to identify a wide range of volatile and semi-volatile low molecular weight compounds.

Metabolomics is highly reliant on the extraction and derivatization steps required for a successful GC-MS-based workflow. Such steps are crucial to ensure the generation of a representative and quantifiable picture of the metabolite profile. The metabolite matrix of legumes like soybean and edamame and their derivative products was inherently complicated and consists of many compounds, including carbohydrates, amino acids, organic acids, fatty acids, polyphenols, and phytochemicals like isoflavones. The chemical properties of these compounds are highly variable (i.e., polarity, volatility, and thermal stability), which requires optimized extraction and derivatization protocols for specific metabolite classes. Polar compounds (e.g., sugar and amino acids) need to be chemically modified (derivatized) to increase their volatility and thermal stability for GC-MS analysis, while the separation of non-polar metabolites (e.g., fatty acids) can be performed with less extensive derivatization.

One well-known and frequently used lipid extraction method is the Bligh-Dyer method, used for gas chromatography-mass spectrometry (GC/MS) analysis. The method of extraction is based on the biphasic solvent system of chloroform, methanol, and water, pioneered by Bligh and Dyer (1959). Some details will be useful to comprehend the essence of this method: The sample is first homogenized in chloroform-methanol; this mixture disrupts cellular structures and releases lipids. Next, the mixture is washed with water, yielding two phases with lipids dissolved in chloroform and non-lipid compounds dissolved in methanol-water. The chloroform layer is then separated, and the lipid fraction is purified for further analysis via gas chromatography-mass spectrometry (GC/MS) [5].

The range of lipid classes separated in the Bligh-Dyer method can also be beneficial (including neutral lipids, phospholipids, and glycolipids) from different biological sources. Such versatility makes it a high-throughput tool for lipidomics research where deep lipid analysis is essential. In contrast, these mild extraction conditions maintain the lipids in their native state and avoid degradation/modification during extraction. It was based on this being demonstrated as a reliable and reproducible method of lipid analyses by both non-MALDI and MALDI based approaches [6], and thus the same principles of point versus the Bligh-Dyer method were largely agreed upon.

In GC/MS application, two common derivatization methods that have been widely used to enhance detection and stability of metabolites, oxime and silylation. The oximation is a reaction of carbonyl compounds with hydroxylamine or derivatives(oxime), producing oximes that are substantially more stable and less volatile than the corresponding parent compounds.

For aldehydes and ketones analysis, derivatization gives important advantages based on their stability and their chromatographic characteristics [7]. In addition, oxime formation increases the molecular mass and introduces specific functional groups that enhance mass spectrometric detection of these compounds and facilitate ionization in the gas phase [7].

In contrast, silylation corresponds to a silyl group introduction on a molecule by reaction, in general with silylating compounds such as N,O-bis(trimethylsilyl)trifluoroacetamide (BSTFA) or N-methyl-N-(trimethylsilyl)trifluoroacetamide (MSTFA) [7]. This standard derivatization method is widely applied to enhance the volatility and thermal stability of polar metabolites, for example, alcohols, amines, and carboxylic acids, for GC/MS identification [8]. Silylation effectively enhances the chromatographic separation and mass spectrometric detection of these compounds by decreasing their polarity and increasing their ionization efficiency [9]. Oximation and silylation together are an effective derivatization approach to analyze a vast set of metabolites for their profiling information that can lead to the understanding of metabolic pathways and biochemical processes [7, 8].

2 Materials

In this section, the reagents and instruments used and the preparation methods in metabolomics profiling of legumes and their derivative products will be described.

2.1 Reagents

The chemicals used in the metabolite profiling of legumes and their derivatives are listed below. This includes solvents, standards, and analytical-grade reagents essential for sample preparation, extraction, and analysis.

1. Liquid nitrogen for rapid quenching.
2. Methanol for solvent-based extraction.
3. Chloroform for solvent-based extraction.
4. Ultrapure water for solvent-based extraction.
5. Ribitol used as an internal standard.
6. Methoxyamine hydrochloride for the derivatization process, oximation.
7. N-methyl-N-(trimethylsilyl)trifluoroacetamide (MSTFA) for derivatization process silylation.

2.2 Instrument

Gas chromatography-mass spectrometry (GC-MS) was employed as the primary analytical tool for the profiling of metabolites in legumes and their derivatives. Nevertheless, the list of instrument usage is as follows.

1. Multi-beads shocker for rapid quenching.
2. Freeze-dryer machine for removing water content while preserving metabolites.
3. Weighing instrument.
4. Thermomixer Incubator for solvent-based extraction and derivatization process.
5. Refrigerated centrifuge for the extraction process, separating debris and liquid, and also hydrophilic and hydrophobic phases.

6. Vacuum evaporator for solvent drying prior to derivatization.
7. Gas chromatography-mass spectrometry (GC-MS).

3 Methods

3.1 Extraction

Extraction process in metabolites profiling of legumes and their derivative product, explained as follows.

3.1.1 Quenching and Pretreatment

To prevent enzymatic activity during extraction, rapid quenching of the sample using liquid nitrogen is essential [10]. Samples that were first quenched and/or frozen at a temperature of −80 °C to ensure rapid solidification of water content. Additionally, homogenization using bead-beating (metal beads) with 10,000 rpm for 10 s enhances metabolite recovery by disrupting cell walls, particularly in non−soy substrates like chickpeas and lentils, which have tougher seed coats compared to soybeans [11]. In Freeze dry treatment, the samples were subjected to primary drying under a vacuum pressure of <0.1 mbar and a chamber temperature of −40 °C to −50 °C for 48–72 h, depending on the sample size and moisture content (*see* **Note 1**).

3.1.2 Solvent-Based Extraction

Methanol-chloroform-water (5:2:2, v/v/v) is a common mixed solvent for the extraction of polar and non-polar metabolites, including amino acids, organic acids, and sugars, from legumes and tempe matrices [12]. Water-soluble metabolites are extracted into the polar phase, and lipophilic compounds are extracted into the non-polar phase. The solvent mixture accommodates the solubilization of hydrophilic compounds with minimization of protein and enzyme denaturation with promising the preservation of metabolites during processing.

1. Weigh the sample for 10 mg in a microtube.
2. A total of 1000 μL of methanol-chloroform-water mixture is added to the sample including internal standard ribitol 50 μg/mL, mixed the solution with vortex for 10 s (*see* **Notes 2** and **3**).
3. Incubate sample using the Thermomixer for 30 min, with 37 °C and 1200 rpm.
4. Centrifuge the sample under cold condition (4 °C, 10,000 rpm, 3 min).
5. Transfer supernatant 500 μL to a new microtube 1.5 mL.
6. Add 100 μL water to the sample to break the equilibrium and separate the hydrophilic and hydrophobic; homogenize it by using a vortex for 10 s.

7. Recentrifuge the sample under the same conditions (4 °C, 10,000 rpm, 3 min).
8. Transfer the upper layer 400 μL to a new microtube and close with a holed caps.
9. Spin dry using vacuum evaporator under room temperature until all the solvent is evaporated for about 1–2 h.

The dried samples were then stored in airtight containers with desiccants at **−40 °C** until further processing, ensuring minimal exposure to moisture and degradation.

3.2 Derivatization

Derivatization is an important step for the analysis of polar, thermally unstable, or nonvolatile metabolites by GC-MS. Specifically, it chemically modifies these metabolites to improve their volatility, thermal stability, and detectability. Sugars, amino acids, and organic acids in legumes, including soybean and edamame, must undergo derivatization before GC-MS can detect them.

Typical derivatizations include silylation, acylation, and alkylation. Of these, silylation with reagents like N,O-bis (trimethylsilyl)trifluoroacetamide (BSTFA) and N-methyl-N-(trimethylsilyl)trifluoroacetamide (MSTFA) is the most widely applied method owing to its ability to effectively transform hydroxyl, carboxyl, and amine to trimethylsilyl derivatives [13]. Legume and derivative product extracts were subjected to a derivatization step before being analyzed with GC/MS as described in the following paragraph.

1. Prepare methoxyamine hydrochloride by weighing and addition with pyridine until the concentration is 20 mg/mL.
2. Add 100 μL methoxyamine hydrochloride solution into the dried sample (*see* **Note 4**).
3. Incubate the sample using a Thermomixer for 90 min under 30 °C, 1200 rpm.
4. Add 50 μL of MSTFA to the sample (*see* **Note 5**).
5. Reincubate the sample for 30 min under 37 °C, 1200 rpm.
6. Centrifuge the sample for 3 min under room condition at 1200 rpm.
7. Transfer 100 μL to the GC/MS vial and the sample ready to be injected.

4 Notes

1. In case of tempe product to prevent stickiness, freeze-dry treatment (overnight) will be performed prior to homogenization by using multi-beads shocker [14].

2. Solid-phase extraction (SPE) is often employed as a secondary purification step to concentrate specific metabolite classes and consider as extraction step. For example, amino acids can be selectively enriched using cation-exchange resin, while phenolic compounds are captured using reverse-phase cartridges [15].
3. For lipophilic metabolites such as fatty acids and sterols, chloroform-methanol (2:1, v/v) is widely utilized in a biphasic extraction approach [16].
4. Acylation, using reagents like acetic anhydride, is often applied for alcohols and amines to improve chromatographic performance and mass spectral ionization [17].
5. Sugars can be derivatized via silylation to yield volatile trimethylsilyl ethers, while fatty acids are commonly analyzed as their methyl ester derivatives (FAMEs) through transesterification [17]. The choice of derivatization reagent is crucial and depends on the metabolite class, target sensitivity, and GC-MS conditions.

References

1. Wei W, Xu Y, Li S, Zhu L, Song J (2015) Developing suppressive soil for root diseases of soybean with continuous long-term cropping of soybean in black soil of Northeast China. Acta Agric Scand Sect B Soil Plant Sci 65:279–285. https://doi.org/10.1080/09064710.2014.992941
2. Messina M, Messina V (2010) The role of soy in vegetarian diets. Nutrients 2(8):855–888. https://doi.org/10.3390/nu2080855
3. Křížová L, Dadáková K, Kašparovská J, Kašparovský T (2019) Isoflavones. Molecules 24(6):1076. https://doi.org/10.3390/molecules24061076
4. Rahmawati D, Astawan M, Putri SP, Fukusaki E (2021) Gas chromatography-mass spectrometry-based metabolite profiling and sensory profile of Indonesian fermented food (Tempe) from various legumes. J Biosci Bioeng 132(5):487–495. https://doi.org/10.1016/j.jbiosc.2021.07.001
5. Sündermann A, Eggers LF, Schwudke D (2016) Liquid extraction: Bligh and Dyer. In: Wenk M (ed) Encyclopedia of lipidomics. Springer, Dordrecht. https://doi.org/10.1007/978-94-007-7864-1_88-1
6. Neag EJ, Collao V, Bhattacharya SK (2023) Analysis of cholesterol lipids using gas chromatography mass spectrometry. In: Bhattacharya SK (ed) Lipidomics. Methods in molecular biology, vol 2625. Humana, New York. https://doi.org/10.1007/978-1-0716-2966-6_13
7. Moldoveanu CS, David V (2019) Derivatization methods in GC and GC/MS. In: Gas chromatography - derivatization, sample preparation, application. IntechOpen. https://doi.org/10.5772/intechopen.81954
8. Djatmika R, Ding W-H (2016) Optimization of silylation for parabens determination by gas chromatography-mass spectrometry. Alchemy 5:55. https://doi.org/10.18860/al.v5i2.3743
9. Khan AI (2014) GC analysis of derivatized amino acids. Thermo Fisher Scientific, Runcorn, Cheshire
10. Chen MM, Li AL, Sun MC, Feng Z, Meng XC, Wang Y (2014) Optimization of the quenching method for metabolomics analysis of Lactobacillus bulgaricus. J Zhejiang Univ Sci B 15(4):333–342. https://doi.org/10.1631/jzus.B1300149. PMID: 24711354; PMCID: PMC3989152.
11. Lee YH, Lee N-R, Lee CH (2022) Comprehensive metabolite profiling of four different beans fermented by aspergillus oryzae. Molecules 27(22):7917. https://doi.org/10.3390/molecules27227917
12. Villaret-Cazadamont J, Poupin N, Tournadre A, Batut A, Gales L, Zalko D, Cabaton NJ, Bellvert F, Bertrand-Michel J (2020) An optimized dual extraction method for the

simultaneous and accurate analysis of polar metabolites and lipids carried out on single biological samples. Metabolites 10(9):338. https://doi.org/10.3390/metabo10090338

13. Villas-Bôas SG, Smart KF, Sivakumaran S, Lane GA (2011) Alkylation or Silylation for analysis of amino and non-amino organic acids by GC-MS? Metabolites 1(1):3–20. https://doi.org/10.3390/metabo1010003. PMID: 24957242; PMCID: PMC4012512.

14. Kadar AD, Aditiawati P, Astawan M, Putri SP, Fukusaki E (2018) Gas chromatography coupled with mass spectrometry-based metabolomics for the classification of Tempe from different regions and production processes in Indonesia. J Biosci Bioeng 126(3):411–416. https://doi.org/10.1016/j.jbiosc.2018.03.020. Epub 2018 Apr 27.

15. Bojko B, Reyes-Garcés N, Bessonneau V, Goryński K, Mousavi F, Souza Silva EA, Pawliszyn J (2014) Solid-phase microextraction in metabolomics. TrAC Trends Anal Chem 61: 168–180. https://doi.org/10.1016/j.trac.2014.07.005

16. Aldana J, Romero-Otero A, Cala MP (2020) Exploring the lipidome: current lipid extraction techniques for mass spectrometry analysis. Metabolites 10(6):231. https://doi.org/10.3390/metabo10060231. PMID: 32503331; PMCID: PMC7345237.

17. Sobolevsky T, Revelsky A, Miller B, Oriedo V, Chernetsova E, Revelsky I (2003) Comparison of silylation and esterification/acylation procedures in GC-MS analysis of amino acids. J Sep Sci 26:1474–1478. https://doi.org/10.1002/jssc.200301492

Chapter 15

GC-MS-Based Metabolite Profiling Method for Thua Nao, a Thai Fermented Soybean Product

Sivamoke Dissook and Romteera Kitichaiworakul

Abstract

Thua Nao, a traditional Thai fermented soybean product, is rich in diverse metabolites contributing to its unique flavor and nutritional profile. Gas chromatography-mass spectrometry (GC-MS) is a powerful analytical tool used to identify and quantify the volatile and semi-volatile compounds in fermented foods like Thua Nao. This chapter provides a detailed methodology for the extraction and derivatization of Thua Nao samples for GC-MS metabolomic analysis. The extraction involves lyophilization, methanol-based solvent extraction, centrifugation, and concentration under vacuum. Derivatization is performed through oximation with methoxyamine hydrochloride, followed by silylation using MSTFA. These steps ensure the effective detection of a broad spectrum of metabolites, including amino acids, organic acids, and volatiles. The protocol optimizes the conditions for sample preparation to improve the reliability and reproducibility of GC-MS analysis in fermented soybean products.

Key words Fermented soybean products, GC-MS, Thua Nao

1 Introduction

Thua Nao is a traditional Thai fermented soybean product widely consumed in northern Thailand [1]. Unlike natto, a well-known Japanese fermented soybean, Thua Nao is nonsticky and is typically sun-dried into sheets or patties. The fermentation process involves *Bacillus subtilis*, which contributes to the product's distinct aroma and metabolic profile [2]. The nutritional and sensory properties of Thua Nao depend on various microbial and enzymatic transformations occurring during fermentation. Thua Nao plays an essential role in northern Thai cuisine as a protein-rich seasoning ingredient. It is often used in stir-fries, curries, and as a component in dipping sauces due to its umami and savory taste. The fermentation process enhances its digestibility and nutritional value by breaking down complex proteins and carbohydrates into more bioavailable forms. This process results in the production of beneficial bioactive

Sastia Prama Putri (ed.), *Gas Chromatography-Mass Spectrometry: Based Metabolomics in Food Science*, Methods and Protocols in Food Science, https://doi.org/10.1007/978-1-0716-5108-7_15,

Fig. 1 Wet Thua Nao. (Source: Personal documentation)

compounds, including those that exhibit potential therapeutic properties. Bioactive compounds derived from the fermentation process have been suggested to exert chemopreventive effects in early-stage colorectal carcinogenesis by modulating gut microbiota and reducing cancer cell proliferation. Additionally, Thua Nao contains isoflavones, such as genistein and daidzein, which have been implicated in cervical cancer treatment by activating the JNK and Erk1/2 MAPK signaling pathways [3, 4].

The fermentation of Thua Nao is traditionally spontaneous, relying on indigenous microbial communities, particularly *Bacillus* spp. This spontaneous fermentation introduces variability in the product's characteristics. Unlike natto, which undergoes controlled fermentation using standardized *B. subtilis* strains, Thua Nao's microbial composition is influenced by environmental factors, raw material quality, and fermentation conditions. Consequently, variations in aroma, texture, and taste are commonly observed among different batches and producers [5]. Thua Nao is available in two primary forms: wet and dried. Wet Thua Nao (Fig. 1) has a strong ammonia-like aroma due to the presence of volatile nitrogenous compounds, which are more pronounced in freshly fermented products. It retains higher moisture content, making it softer and more perishable. Traditionally, wet Thua Nao is used immediately after fermentation, often mixed into dishes or mashed into a paste with additional seasonings such as garlic, onion, and chili [6].

Dried Thua Nao (Fig. 2), on the other hand, undergoes a sun-drying process that reduces its moisture content significantly. This form is commonly molded into thin, circular disks and dried in the sun for one or more days. Drying not only extends its shelf life but also moderates the strong ammonia odor found in wet Thua

Fig. 2 Dried Thua Nao. (Source: Personal documentation)

Nao. The drying process causes changes in its volatile profile, leading to a reduction in ammonia and an increase in nutty and roasted notes, primarily due to the concentration of pyrazines and Maillard reaction products [7]. Although dried Thua Nao has a milder aroma, it still retains its characteristic umami and fermented flavors, making it a preferred form for long-term storage and transport. Both forms of Thua Nao serve distinct culinary purposes. Wet Thua Nao is often incorporated into fresh dishes where a more intense fermented taste is desirable, whereas dried Thua Nao is used as a flavor enhancer, ground into powder, or rehydrated before cooking. The choice between wet and dried forms depends on the desired intensity of flavor and the method of preparation [1].

This chapter will introduce a method for extracting metabolites from both wet and dried Thua Nao, (*see* **Note 1**) ensuring the reliable identification of key flavor and bioactive compounds. The complexity of Thua Nao's biochemical matrix necessitates a method that optimizes the extraction of volatile and semi-volatile compounds while minimizing degradation.

2 Materials

2.1 Reagents for Wet Thua Nao Preparation

The reagents for wet Thua Nao preparation are listed in Table 1.

2.2 Reagents for Dried Thua Nao Preparation

The reagents for dried Thua Nao preparation are listed in Table 2.

Table 1
Wet Thua Nao preparation reagents

Purpose:	Extracts both polar (organic acids, sugars) and nonpolar (lipids, pyrazines) compounds while preventing enzymatic degradation
Solvents and reagents:	1. Methanol (LC-MS grade) 2. Chloroform 3. Milli-Q water
Internal standard:	1. Ribitol 2. Heptadecanoic acid
Derivatization reagents:	1. Methoxyamine hydrochloride (20 mg/mL in pyridine) 2. MSTFA (N-methyl-N-(trimethylsilyl)trifluoroace-tamide)
Equipment:	1. Sonicator 2. Centrifuge 3. Nitrogen evaporator 4. Thermomixer 5. GC-MS vials

Table 2
Dried Thua Nao preparation reagents

Purpose:	Extracts Maillard products, pyrazines, fatty acids, and aldehydes while minimizing oxidation
Solvents and reagents:	1. Isopropanol 2. Milli-Q water
Internal standard:	1. Heptadecanoic acid 2. Phenylacetone
Derivatization reagents:	1. Methoxyamine hydrochloride (20 mg/mL in pyridine) 2. MSTFA (N-methyl-N-(trimethylsilyl)trifluoroacetamide)
Equipment:	1. Bead mill grinder 2. Sonicator 3. Centrifuge 4. Nitrogen evaporator 5. Thermomixer 6. GC-MS vials

3 Methods

3.1 Wet Thua Nao Extraction and Derivatization Method

Table 3 shows the extraction and derivatization procedures for wet Thua Nao.

3.2 Dried Thua Nao Extraction and Derivatization Method

The extraction and derivatization procedures for dried Thua Nao are included in Table 4.

Table 3
Wet Thua Nao extraction and derivatization procedures

1. *Sample preparation:*	Weigh 200 mg of wet Thua Nao into a 2 mL Eppendorf tube
2. *Quenching and homogenization:*	Add 1 mL of cold methanol: water (3:1 v/v) to stop enzymatic activity Vortex for 30 s
3. *Lipid separation:*	Add 500 μL of chloroform Vortex for 30 s Sonicate for10 min
4. *Centrifugation:*	Centrifuge at 12,000× *g*, 4 °C for 10 min Collect the upper (methanol-water) phase for sugars, organic acids. Collect the lower (chloroform) phase for lipids, fatty acids.
5. *Drying:*	Evaporate solvent under nitrogen at 35 °C
6. *Derivatization:*	Add 50 μL of methoxyamine hydrochloride (20 mg/mL in pyridine) Incubate at 60 °C for 90 min at 1200 rpm (*see* **Notes 2** and **3**) Add 100 μL of MSTFA Incubate at 70 °C for 30 min at 1200 rpm to form TMS derivatives
7. *GC-MS analysis:*	Transfer to GC vial, cap, and analyze.

Table 4
Dried Thua Nao extraction and derivatization procedures

1. *Sample preparation:*	Weigh 100 mg of dried Thua Nao, grind into fine powder using a bead mill
2. *Rehydration:*	Add 1 mL of isopropanol: water (1:1 v/v) Vortex for 30 s
3. *Sonication:*	Ultrasonicate for 10 min at 4 °C to enhance extraction
4. *Centrifugation:*	Centrifuge at 12,000× *g*, 4 °C for 15 min
5. *Supernatant collection:*	Transfer liquid phase to a new vial
6. *Drying:*	Evaporate solvent under nitrogen at 35 °C
7. *Derivatization:*	Same as wet Thua Nao protocol (methoxyamine + MSTFA)
8. *GC-MS analysis:*	Transfer to GC vial, cap, and analyze

3.3 Operating Conditions

The GC/MS operating conditions for Thua Nao analysis are provided in Table 5. Adjustments may be necessary if different conditions are applied.

4 Notes

1. Comparison of extraction methods (Table 6).
2. For reliable GC-MS analysis of fermented soybean metabolites, methoxyamine (MOX) derivatization at 60 °C for 90 min

Table 5
GC/MS operating conditions for Thua Nao analysis

Column:	DB-5MS (30 m × 0.25 mm × 0.25 μm)
Carrier gas:	Helium (1 mL/min flow rate)
Injection:	Splitless mode, 1 μL of injection
Oven program:	70 °C (2 min hold) Ramp 10 °C/min to 250 °C Hold at 250 °C for 10 min
Detector:	Mass spectrometer (EI mode, 70 eV)
Mass range:	40–600 m/z

Table 6
Comparison of different extraction methods for Thua Nao analysis

Factor	Wet Thua Nao (Methanol-Chloroform)	Dried Thua Nao (IPA Extraction)
Best for	Amino acids, organic acids, sugars, lipids	Pyrazines, Maillard reaction products, aldehydes
Solvent	Methanol:Water:Chloroform	Isopropanol:Water
Derivatization	Methoxyamine + MSTFA	Methoxyamine + MSTFA
Enzyme inactivation	Yes (quenching with cold solvent)	Not required (enzymes inactive)

ensures full stabilization of reducing sugars, organic acids, and keto acids, preventing peak splitting and enhancing quantification accuracy. This approach balances efficiency and metabolite stability, minimizing degradation of heat-sensitive compounds like pyruvic acid and amino acids.

3. For applications requiring faster turnaround, 70 °C for 30 min offers a viable alternative, though optimization is recommended to prevent thermal degradation.

References

1. Dajanta K, Apichartsrangkoon A, Chukeatirote E, Frazier RA (2011) Volatile profiles of *Thua Nao*, a Thai fermented soy product. Food Chem 125(2):464–470. https://doi.org/10.1016/j.foodchem.2010.09.030
2. Kai M (2020) Diversity and distribution of volatile secondary metabolites throughout *Bacillus subtilis* isolates. Front Microbiol 11:559. https://doi.org/10.3389/fmicb.2020.00559
3. Taya S, Dissook S, Ruangsuriya J, Yodkeeree S, Boonyapranai K, Chewonarin T, Wongpoomchai R (2024) Thai fermented soybean (*Thua-Nao*) prevents early stages of colorectal carcinogenesis induced by diethylnitrosamine and 1,2-dimethylhydrazine through modulations of cell proliferation and gut microbiota in rats. Nutrients 16:3506. https://doi.org/10.3390/nu16203506

4. Sukhamwang A, Inthanon S, Dejkriengkraikul P, Semangoen T, Yodkeeree S (2024) Anti-cancer potential of isoflavone-enriched fraction from traditional Thai fermented soybean against Hela cervical cancer cells. Int J Mol Sci 25: 9277. https://doi.org/10.3390/ijms25179277
5. Visessanguan W, Benjakul S, Potachareon W, Panya A, Riebroy S (2005) Accelerated proteolysis of soy proteins during fermentation of *Thua Nao* inoculated with *Bacillus subtilis*. J Food Biochem 29(4):349–366. https://doi.org/10.1111/j.1745-4514.2005.00012.x
6. Leejeerajumnean A, Duckham SC, Owens JD, Ames JM (2001) Volatile compounds in *Bacillus*-fermented soybeans. J Sci Food Agric 81(5): 525–529. https://doi.org/10.1002/jsfa.843
7. Mathatheeranan P, Wongprasert T, Nan TN, Suwannakul E, Wang Y, Lu TJ et al (2023) Volatile profiles and aroma-active components of Northern Thai fermented soybean (*Thua Nao*) produced from controlled and uncontrolled fermentations. Int J Gastron Food Sci 33:100763. https://doi.org/10.1016/j.ijgfs.2023.100763

Chapter 16

GC-MS-Based Metabolite Profiling Method for Soy Sauce

Asuka Mori and Yoshika Maekawa

Abstract

Soy sauce is a fermented seasoning with a strong umami, salty taste, and unique aroma. It is brewed through a two-step fermentation process called koji fermentation and moromi fermentation. The flavor of soy sauce is formed through complex microbial interactions. Gas chromatography-mass spectrometry (GC-MS)-based metabolomics is a widely used method for analyzing flavor-related compounds in soy sauce. This chapter summarizes the protocol for analyzing both volatile and nonvolatile compounds using GC-MS, from sample preparation to GC-MS analytical conditions.

Key words Soy sauce, GC-MS, Metabolomics, Volatile, Nonvolatile

1 Introduction

Soy sauce is a traditional seasoning brewed from soybeans, wheat, and salt by koji fermentation and moromi fermentation [1, 2]. Koji fermentation is a solid-state fermentation of soybeans and wheat by koji mold, typically *Aspergillus oryzae* or *Aspergillus soja* [1, 3]. During this step, the enzymes produced from the koji mold break down the raw materials into peptides, amino acids, and saccharides [3, 4]. The second stage is moromi fermentation. Koji is added to a brine solution containing 18–22% NaCl, resulting in the growth of indigenous halotolerant species, including lactic acid bacteria and yeast [3, 4]. These microorganisms utilize the peptides, amino acids, and saccharides produced in the koji fermentation as nutrient sources for growth [1]. During this two-step fermentation process, the flavor of the soy sauce is formed through complex microbial interactions [4]. Based on the amount of wheat used for production, soy sauce can be classified into Japanese-type and Chinese-type [5]. Japanese-type uses equal amounts of soybeans and wheat, while Chinese-type uses mainly soybeans rather than wheat [1].

Sastia Prama Putri (ed.), *Gas Chromatography-Mass Spectrometry: Based Metabolomics in Food Science*, Methods and Protocols in Food Science, https://doi.org/10.1007/978-1-0716-5108-7_16,

The flavor of soy sauce is characterized by its salty and umami taste, as well as its caramel-like aroma [1]. However, various compounds are formed during the complex microbial interaction. In addition, the differences in the fermentation conditions and the ratio of ingredients result in variations in the final flavor of soy sauce. Therefore, it is important to comprehensively analyze the flavor-related compounds in soy sauce for quality improvement in the final products. Gas chromatography-mass spectrometry (GC-MS)-based metabolomics is often utilized to analyze both nonvolatile (taste-related) compounds and volatile (aroma-related) compounds in soy sauce [5–11]. This chapter summarizes the detailed procedure for GC-MS analysis of soy sauce, including sample preparation, extraction, and GC-MS parameters.

2 Materials

2.1 Nonvolatile Metabolite Analysis

2.1.1 Samples and Reagents

1. Soy sauce.
2. Liquid nitrogen.
3. Ultrapure water.
4. DL-alanine-2,3,3,3-d_4 (as internal standard (IS)).
5. Methoxyamine hydrochloride.
6. Infinity Pure pyridine (99.5 + %).
7. *N*-methyl-*N*-(trimethylsilyl)-trifluoroacetamide (MSTFA).
8. C_9–C_{40} alkane standard mixture.

2.1.2 Laboratory Supplies

1. Micropipette.
2. Freezer gloves for handling frozen objects.
3. 4 °C refrigerator to store samples.
4. 2.0 mL microcentrifuge tubes.
5. 11 mL vial (for storing soy sauce samples).
6. Vortex mixer.
7. Thermomixer.
8. Sonicator.
9. GC-MS vials equipped with 200 μL insert.

2.1.3 GC-MS Equipment

1. GC-MS system: GCMS-QP2010 Ultra (Shimadzu) equipped with AOC-20i/s auto-injector (Shimadzu) or equivalent system.
2. GC capillary column: InertCap 5 MS/NP column (GL Sciences) or equivalent column.
3. NIST20 MS Spectral Library (National Institute of Standards and Technology).

4. GL-Science DB (InertCap 5MS Metabolomics, Kovats RI)—can be downloaded on the MS-DIAL website.

2.2 Volatile Metabolite Analysis

2.2.1 Samples and Reagents

1. Soy sauce.
2. Fatty acid ethyl esters (FAEEs, C_4–C_{24}).
3. NaCl.

2.2.2 Laboratory Supplies

1. SPME vial (20 mL screw-cap vial).
2. 100 mL screw vial (for QC sample).
3. Micropipette.

2.2.3 GC-MS Equipment

1. GC-MS system: GCMS-TQ8050 NX (Shimadzu) equipped with AOC-6000 autosampler (Shimadzu).
2. SPME Arrow: DVB/CAR/PDMS (Divinylbenzene/Carboxen/Polydimethylsiloxane) fiber (Shimadzu).
3. GC capillary column: InertCap FFAP (0.32 mm × 60 m, 0.50 μm; GL Sciences).
4. NIST20 MS Spectral Library (National Institute of Standards and Technology).

3 Methods

3.1 Nonvolatile Metabolite Analysis

3.1.1 Sample Preparation and Extraction

1. Aliquot the soy sauce samples into 11 mL vials and store them in a refrigerator at 4 °C until analysis (*see* **Note 1**).
2. Prepare IS-containing ultrapure water consisting of IS (DL-alanine-2,3,3,3-d_4 200 μg/mL) and ultrapure water.
3. Transfer 20 μL of the soy sauce sample into a 2.0 mL microcentrifuge tube. Prepare microcentrifuge tubes for blank samples as well. For blank samples, use 20 μL of ultrapure water instead of the soy sauce sample.
4. Add 180 μL of IS-containing ultrapure water to dilute the sample and vortex.
5. Transfer 20 μL of the supernatant to a new 2.0 mL microcentrifuge tube. Collect 20 μL of supernatant from each sample for quality control (QC) samples into a 2.0 mL microcentrifuge tube, and then vortex.
6. Transfer 20 μL of QC sample into a new 2.0 mL microcentrifuge tube.
7. Close all tubes with perforated caps.
8. Freeze all tubes in liquid nitrogen.
9. Lyophilize overnight (*see* **Note 2**).

3.1.2 Derivatization

1. Freshly prepare methoxyamine hydrochloride (20 mg/mL) in pyridine and mix vigorously until dissolved.
2. Add 50 μL of methoxyamine hydrochloride solvent to the lyophilized samples and sonicate for 1 min.
3. Incubate the samples in a thermomixer at 30 °C for 90 min and 1200 rpm.
4. Add 50 μL of MSTFA to each sample.
5. Incubate the resulting solution at 37 °C for 30 min and 1200 rpm.
6. Transfer 50 μL of the supernatant into a GC-MS vial and cap firmly. Samples are ready to inject.

3.1.3 GC-MS Parameters

1. The injection amount is 1 μL with a split ratio of 1:25.
2. The injection temperature is set to 270 °C.
3. Set the helium carrier gas flow rate to 1.12 mL/min.
4. The column temperature is initially set at 80 °C for 2 min, then increased by 15 °C/min until reaching the final temperature of 330 °C—hold for 6 min.
5. The transfer line temperature is set to 310 °C. The ion source temperature is set at 280 °C.
6. The electron ionization operates at 70 eV.
7. Obtain the mass spectra in the scan range of 85–500 *m/z*. Event time is 0.100 s.
8. A standard alkane mixture (C_9–C_{40}) is set to be injected first before the samples.

3.2 Volatile Metabolite Analysis

3.2.1 Sample Preparation and Extraction

1. Weigh 3 g of NaCl and place it in a 20 mL SPME vial.
2. Add 6 mL of the soy sauce sample into the vial (*see* **Note 3**).
3. For QC sample, pool 2.0 mL from all soy sauce samples into a 100 mL screw vial and mix well using a vortex mixer. Put 6 mL of QC sample into a 20 mL SPME vial.
4. Sample extraction using SPME is performed by an AOC-6000 autosampler. Before sample extraction, the DVB/CAR/PDMS fiber is pre-conditioned at 250 °C for 10 min. Incubate the sample vial at 50 °C for 10 min with shaking.
5. Expose the fiber to the samples at 50 °C for 30 min with agitation (250 rpm) to extract the volatile compounds from samples.
6. Following the extraction, the analyte is injected via thermal desorption at 250 °C for 2 min.

3.2.2 GC-MS Parameters

1. The analyte desorbed from the SPME fibers is injected in splitless mode.
2. A helium carrier gas with a linear velocity of 40.7 cm/s.
3. The column temperature is initially set at 40 °C for 3 min, then increased by 3 °C/min until reaching the final temperature of 250 °C and maintained at 250 °C for 45 min.
4. The MS is operated in scan mode (*m/z* 24–350). Event time is 0.100 s.
5. The ion source and interface temperatures are set at 250 °C.
6. Fatty acid ethyl esters (C_4–C_{24}) are injected at the beginning of the analysis for the calculation of retention index.

4 Notes

1. Prepare a soy sauce sample in an 11 mL vial, avoiding as much air as possible in the headspace because the soy sauce will be oxidized.
2. Cut the lid of the microcentrifuge tube with scissors and drill three holes in the lid. Insert the tube into the floater for lyophilization.
3. When putting the soy sauce in the vial, make sure it does not adhere to the walls.

References

1. Diez-Simon C, Eichelsheim C, Mumm R, Hall RD (2020) Chemical and sensory characteristics of soy sauce: a review. J Agric Food Chem 68:11612–11630. https://doi.org/10.1021/acs.jafc.0c04274
2. Chen X-C, Meng F-B, Tian H-L et al (2024) Dynamic changes in metabolomics and flavoromics during industrial scale fermentation of Chinese traditional soy sauce. Food Biosci 59: 103940. https://doi.org/10.1016/j.fbio.2024.103940
3. Ito K, Matsuyama A (2021) Koji molds for Japanese soy sauce brewing: characteristics and key enzymes. J Fungi 7:658. https://doi.org/10.3390/jof7080658
4. Devanthi PVP, Gkatzionis K (2019) Soy sauce fermentation: microorganisms, aroma formation, and process modification. Food Res Int 120:364–374
5. Wanakhachornkrai P, Lertsiri S (2003) Comparison of determination method for volatile compounds in Thai soy sauce. Food Chem 83:619–629
6. Harada R, Yuzuki M, Ito K et al (2018) Microbe participation in aroma production during soy sauce fermentation. J Biosci Bioeng 125:688–694
7. Harada R, Yuzuki M, Ito K et al (2017) Influence of yeast and lactic acid bacterium on the constituent profile of soy sauce during fermentation. J Biosci Bioeng 123:203–208
8. Shiga K, Yamamoto S, Nakajima A et al (2014) Metabolic profiling approach to explore compounds related to the umami intensity of soy sauce. J Agric Food Chem 62:7317–7322. https://doi.org/10.1021/jf501173r
9. Yamamoto S, Shiga K, Kodama Y et al (2014) Analysis of the correlation between dipeptides and taste differences among soy sauces by using metabolomics-based component profiling. J Biosci Bioeng 118:56–63

10. Yamamoto S, Bamba T, Sano A et al (2012) Metabolite profiling of soy sauce using gas chromatography with time-of-flight mass spectrometry and analysis of correlation with quantitative descriptive analysis. J Biosci Bioeng 114:170–175
11. Reddy TR, Overmyer KA, Coon JJ et al (2021) Metabolomic biomarkers differentiate soy sauce freshness under conditions of accelerated storage. J Food Qual 2021:1–11. https://doi.org/10.1155/2021/6650990

Chapter 17

GC-MS-Based Metabolite Profiling Method for Miso

Chiaki Tsushi and Hiroki Naito

Abstract

Miso is a traditional fermented food that represents Japan, and its unique flavor and functionality are due to the presence of various metabolites. In this chapter, we introduce the gas chromatography-mass spectrometry (GC-MS) analysis method with the aim of comprehensively clarifying the overall picture of the nonvolatile flavor components that make up miso. GC-MS can be used to analyze nonvolatile components through derivatization and to extract molecules that contribute to the aroma of miso. This method enables the identification and profiling of metabolites characteristic of miso. This chapter describes in detail the pretreatment and analysis procedures for metabolomic analysis of miso using GC-MS and is expected to serve as a useful guideline for conducting similar research in the future.

Key words GC-MS, Metabolites, Miso, Nonvolatile

1 Introduction

Miso is a traditional fermented food in Japan and has long contributed to the nutrition and health maintenance of Japanese people as a source of high-quality vegetable protein. Miso is mainly made from soybeans, rice, and salt and is produced through a fermentation process at room temperature for several months to more than 24 months [1]. A variety of microorganisms, such as koji mold, lactic acid bacteria, and yeast are involved in the fermentation process, and the balance and interaction of these microbial communities greatly contribute to the formation of miso's unique flavor and aroma, as well as its nutritional value and functional components. In addition, it is known that various variations in the quality and characteristics of miso can be created depending on the type and composition ratio of soybeans, rice, and salt, which are used as raw materials, and differences in fermentation conditions [2].

Recent studies have revealed that miso has a variety of health functions, including anticancer, antibacterial, and anti-obesity effects [3]. In particular, it has been clarified that functional

Sastia Prama Putri (ed.), *Gas Chromatography-Mass Spectrometry: Based Metabolomics in Food Science*, Methods and Protocols in Food Science, https://doi.org/10.1007/978-1-0716-5108-7_17,

components, such as furazan, which has anti-fat-producing and antioxidant effects, are present in miso, further enhancing the value of miso as a health-maintaining food [4].

In this way, since the quality and functionality of miso are greatly affected by multiple factors such as the selection of raw materials, the blending ratio, the diversity of microorganisms involved in fermentation, and the fermentation conditions, it is extremely important to comprehensively evaluate the quality of miso in order to ensure the stability of the product, develop a variety of products according to consumer preferences, and improve health functionality.

In the quality assessment of miso, critical attributes such as aroma and flavor are predominantly evaluated using sensory analysis by human assessors [5]. Since the outcomes of sensory evaluations can be influenced by the individual characteristics of the assessors, it is recommended that such evaluations be conducted by personnel who have received appropriate training and possess considerable experience.

In recent years, metabolomics—a comprehensive approach to profiling metabolic components—has gained significant attention. When applied to food systems, this technique enables the simultaneous analysis of chemical compounds that contribute to sensory properties such as taste and aroma, thereby providing objective data free from human subjectivity [6]. Various analytical methods have been employed for these purposes [7, 8]. For example, GC-MS analysis of hydrophilic compounds in miso has demonstrated a correlation between galactose and umami taste [9].

Among the analytical methods commonly employed in metabolomics, GC-MS offer a significant advantage due to the extensive compound libraries available. This abundance facilitates relatively straightforward compound identification, thus lowering the entry barrier compared to other techniques. However, regardless of the method chosen, the types and numbers of detected compounds can vary considerably depending on the sample pretreatment procedures, such as extraction [10]. Moreover, published literature often lacks sufficient details regarding experimental procedures and instrumentation, underscoring the need for comprehensive protocols in the pretreatment and analysis of miso.

In this chapter, we provide a detailed description of the sample pretreatment and analytical workflows for miso metabolomics using GC-MS. The objective is to establish a methodological framework that researchers can adopt as a foundation, thereby facilitating the acquisition of robust and reproducible data.

2 Materials

2.1 Samples and Reagents

1. Miso.
2. Ribitol.
3. Ultrapure water.
4. Methanol.
5. Chloroform.
6. N-methyl-N-(trimethylsilyl) trifluoroacetamide (MSTFA).
7. Liquid nitrogen.
8. Alkane mix($C_{10} \sim C_{38}$).

2.2 Laboratory Supplies

1. Multi-beads shocker.
2. 50 mL cryogenic grinding tube.
3. 2 mL microcentrifuge tubes.
4. 1.5 mL microcentrifuge tubes.
5. Vortex mixer.
6. volumetric pipette.
7. Warming shaker.
8. Refrigerated centrifuge.
9. Compact benchtop centrifuge.
10. Micropipettes.
11. Centrifugal evaporator.
12. Freeze dryer.
13. GC-MS vial.

2.3 GC-MS Equipment

1. GC-MS system: GCMS-QP2010 Ultra (Shimadzu) or equivalent system.
2. Auto-sampler: The AOC-20i/s (Shimadzu).
3. Column: The InertCap 5MS/NP (0.25 mm × 30 m, 0.25 μm, GL Sciences) or equivalent system.

3 Methods

3.1 Sample Preparation

1. Weigh 5.0 g of the miso sample (see Note 1) into a cryogenic grinding tube and immediately freeze it using liquid nitrogen.
2. Add metal beads to the tube and pulverize the frozen sample using a multi-beads shocker. (see Note 2)
3. Weigh 20 mg of ribitol (which serves as the internal standard) into a 1.5 mL tube, then add 1 mL of ultrapure water.

4. To perform extraction via the Bligh-Dyer method, use a volumetric pipette to prepare a mixed solvent of ultrapure water, methanol, and chloroform in a 2:5:2 ratio. The total volume of the solvent should be calculated as (number of samples × 900 μL + α).

3.2 Extraction of Hydrophilic Metabolites

1. Weigh 10 mg of the pulverized sample into a 2 mL tube. (see Note 3)
2. Add 1.0 mL of the prepared mixed solvent (ultrapure water: methanol: chloroform = 2:5:2) to the sample and mix thoroughly using a vortex mixer.
3. Place the sample on a warm shaker set at 37 °C and agitate at 1200 rpm for 30 min.
4. Centrifuge the sample at 9390 g for 5 min at 4 °C. (see Note 4)
5. Transfer 800 μL of the resulting supernatant to a new 2 mL tube, add 500 μL of distilled water, and mix using a vortex mixer.
6. Transfer 100 μL of the supernatant to a separate 1.5 mL tube, perform centrifugal concentration for 30 min, and subsequently freeze-dry the sample overnight.

3.3 Derivatization

1. Prepare a 20 mg/mL solution of methoxyamine hydrochloride in pyridine. The total volume should be calculated as (number of samples × 900 μL + α), where α accounts for additional volume to compensate for pipetting losses.
2. To each dried sample (see Note 5), add 100 μL of the methoxyamine solution and incubate on a warm shaker at 30 °C for 90 min.
3. Subsequently, add 50 μL of MSTFA to the sample and incubate on a warm shaker at 37 °C for 30 min.
4. Transfer 100 μL of the resulting supernatant into a GC injection vial.
5. Verify that no air bubbles are present at the bottom of the vial.

3.4 GC Conditions

1. Injection temperature: 230 °C.
2. Split mode: 50:1.
3. Carrier gas: He (flow rate: 1.12 mL/min).
4. Column temperature program: Hold at 80 °C for 2 min, then ramp at 15 °C/min to 330 °C and hold for 6 min.

3.5 MS Conditions

1. Ion source temperature: 200 °C.
2. Interface temperature: 250 °C.
3. Scan speed: 10,000 u/s.

4. Scan range: m/z 85–500.
5. EI voltage: 70 kV.

4 Notes

1. *Storage conditions*: To prevent the fermentation of miso and changes in metabolites during the time until analysis, storage in a freezer at −30 °C or below is recommended.
2. *Challenges in freeze pulverization*: In freeze pulverization, depending on the type of miso, there may be a high sugar content, making it difficult to powder. In such cases, extending the freezing time or using freeze-drying can be effective.
3. *Temperature when weighing the sample*: Please wait for the frozen powdered sample to reach room temperature before weighing it. This is to avoid moisture absorption by the sample due to condensation from temperature changes.
4. *Temperature management*: Some types of miso contain a high amount of enzymes, which may lead to changes in metabolites during the extraction process. Therefore, low temperatures (4 °C) should be used during centrifugation and other operations.
5. *Drying before derivatization*: Ensure that drying is sufficiently performed before derivatization. Depending on the performance of the equipment and the sample, the overnight freeze-drying time can be shortened. However, conducting derivatization with residual moisture can reduce derivatization efficiency and lead to a decrease in the quality and stability of the spectrum.

References

1. Qin P, Wang T, Luo Y (2022) A review on plant-based proteins from soybean: health benefits and soy product development. J Agric Food Res 29(4):277–288. https://doi.org/10.1016/j.jafr.2021.100265
2. Allwood JG, Wakeling LT, Bean DC (2021) Fermentation and the microbial community of Japanese koji and miso: a review. J Food Sci 86(6):2194–2207. https://doi.org/10.1111/1750-3841.15773
3. Saeed F, Afzar M, Shah YA, Khan MH, Hussain M, Ikram A, Atiq H, Norman M, Sewan SA, Hashroom AO (2022) Miso: a traditional nutritious & health-endorsing fermented product. Food Sci Nutr 10(12):4103–4111. https://doi.org/10.1002/fsn3.3029
4. Li Y-J, Zhang Y-P, Chiu P-C, Lin S-M, Lee C, Lo C-Y, Lien W-T, Chiou RY-Y (2022) Presence of Flazin in miso and identification of four tryptophan-derived ß-Carbolines formed during fermentation. J Agric Food Chem 72(38):21065–21076. https://doi.org/10.1021/acs.jafc.4c05855.s001
5. Wang S, Tamura T, Kyouno N, Liu X, Zhang H, Akiyama Y, Yu Chen J (2019) Effect of the chemical composition of miso (Japanese fermented soybean paste) upon the sensory evaluation. Anal Lett 52(11):1813–1827. https://doi.org/10.1080/00032719.2019.1570244

6. Shu N, Chen X, Sun X, Cao X, Liu Y, Xu Y-J (2023) Metabolomics identify landscape of food sensory properties. Crit Rev Food Sci Nutr 63(27):8478–8488. https://doi.org/10.1080/10408398.2022.2062698
7. Taheri S, de Andrade JC, Conte-Junior CA (2024) Emerging perspectives on analytical techniques and machine learning for food metabolomics in the era of industry 4.0: a systematic review. Crit Rev Food Sci Nutr:1–27. https://doi.org/10.1080/10408398.2024.2435597
8. Segers K, Declerck S, Mangelings D, Heyden YV, Van Eeckhaut A (2019) Analytical techniques for metabolomic studies: a review. Bioanalysis 11(24):2297–2318. https://doi.org/10.4155/bio-2019-0014
9. Ogawa T, Kusumoto K-I, Fukusaki E (2019) Construction of a prediction model for taste of miso (Japanese fermented soybean paste) using metabolic profiling and quantitative descriptive analyses. Food Sci Technol Res 25(6):871–877. https://doi.org/10.3136/fstr.25.871
10. Mushtaq MY, Choi YH, Verpoorte R, Wilson EG (2014) Extraction for metabolomics: access to the metabolome: extraction for metabolomics: access to the metabolome. Phytochem Anal 25(4):291–306. https://doi.org/10.1002/pca.2505

Part IV

Aquaculture Products

Chapter 18

GC-MS-Based Metabolite Profiling Method for Fish

Ryota Mabuchi

Abstract

This chapter discusses the extraction and derivatization methods used for the metabolomic analysis of fish samples via GC-MS. It is essential to select an extraction method based on the research objectives.

Key words Ethanol or water extraction, Extractive components, Fish meat, GC-MS metabolomics, Methanol/chloroform/ultrapure water, Oxime-trimethylsilyl, Primary metabolites, Taste substance, Water-soluble low-molecular-weight components

1 Introduction

In metabolomics research focused on food science, GC-MS targeting fish has been conducted mainly on primary metabolites, odor components, and fatty acids. The most common type of research targets water-soluble, low-molecular-weight components, such as sugars, amino acids, and organic acids. GC-MS is used to analyze changes in the freshness and degree of decomposition of fish, using water-soluble, low-molecular-weight components as indicators [1]. Water-soluble, low-molecular-weight components change during the metabolism of fish, and many of them can be tasted by humans. Therefore, GC-MS can assess the deliciousness of fish using taste substances as indicators [2].

Similar to other foods, metabolomic research is also being conducted on odorous components. In the case of fish, many odor components are produced as the fish deteriorates and spoils, leading to a decline in its quality [3].

Furthermore, lipidomics research targeting fatty acids has been conducted, although it is included in the broader definition of metabolomics [4]. Fatty acids in fish are also related to their deliciousness; therefore, fatty acid analysis is essential in the field of food science research on fish.

Sastia Prama Putri (ed.), *Gas Chromatography-Mass Spectrometry: Based Metabolomics in Food Science*, Methods and Protocols in Food Science, https://doi.org/10.1007/978-1-0716-5108-7_18,

Our research group used a metabolomic approach on water-soluble, low-molecular-weight components. We applied analytical data, mainly component indices, to evaluate fish quality. For example, we are working on the identification of fish meat parts [5], the construction of new freshness evaluation indices [1], and their application in the evaluation of fish flavor [2, 6]. We also searched for heating history markers based on changes in the components that occur when fish meat is heated [7].

This chapter focuses on the metabolomic methods used for water-soluble, low-molecular-weight components and introduces the protocols for the extraction and derivatization methods. The extraction method is crucial in fish metabolomic research.

Traditionally, research on the quality and flavor of fish has focused on extractive components (nitrogenous extractive components). Although the terms "water-soluble low molecular weight components" and "nitrogenous extractive components" are almost synonymous, the author uses the term "water-soluble low molecular weight components" when referring to nontargeted analysis, such as that carried out in the early stages of metabolomics research. However, in the case of target analysis, where the focus is on specific components, it is easier to understand whether we can distinguish between the two by referring to them as nitrogenous extractive components.

Various methods have been devised for extracting the extractive components, and it has been reported that the obtained extractive components differ depending on the extraction method used [8]. Typical extraction methods involve the use of perchloric acid (PCA), trichloroacetic acid (TCA), ethanol, and water. In addition, extraction was performed using picric and tungstic acids. PCA has a relatively high extraction efficiency for free amino acids. Water or ethanol extraction (solvents that are safe for humans) is generally used for sensory evaluation. Previous studies have evaluated the taste of extracted components based on the sensory values obtained from sensory evaluation and the component amounts obtained from component analysis [9].

There are various methods for extracting the components, and it is recommended to choose the method most appropriate for the experiment. Therefore, the question arises: Which extraction method is best for metabolomic research using GC-MS? There are two primary approaches to this problem.

First, it is necessary to accumulate component data obtained through metabolomic research. In the field of food science, the method using a solvent with a composition of methanol/chloroform/water (2.5/1/1 = v/v/v) is the mainstream method for metabolomics research using GC-MS [10]. This extraction method is also being used in the fish metabolomics research conducted by the author. This extraction method made it possible to interpret the results in a cross-sectional manner with other experiments and accumulate experimental data. Second, when considering food, research work should be carried out assuming that the food will actually be eaten. We are currently attempting to evaluate the deliciousness of fish by comparing the amount of taste components in fish extracts and sensory values using GC-MS metabolomics. In this study, it was preferable to use an extraction method that assumed that the food would be consumed by humans. For this reason, our laboratory uses methods that use water and ethanol.

Therefore, this chapter will include protocols for the most commonly used extraction methods, which use methanol/chloroform/water, water-only, and 80% ethanol. In addition, the differences in the component profiles obtained using these extraction methods are visualized and shown in the "Notes" (*see* **Note 1**).

However, there are no particular points to note when using the derivatization method in experiments on fish; therefore, there are no problems with using the same derivatization method for other foods. *N*-Methyl-*N*-trimethylsilyltrifluoroacetamide (MSTFA) has been used in various GC-MS metabolomic studies, and our laboratory uses this reagent [10]. Studies using TBDMS and BSTFA have also been reported. However, MSTFA has been used in many studies, and we believe that it is better to use the same reagent to accumulate data. Therefore, the derivatization methods described in this chapter introduce only general derivatization methods (oximation and silylation using MSTFA).

2 Materials

1. Methanol/chloroform/ultrapure water (2.5/1/1 = v/v/v) (for extracting water-soluble, low-molecular-weight components).
2. Ultrapure water (used for water extraction).
3. Ethanol (diluted to 80% and used for 80% ethanol extraction).
4. Ribitol in ultrapure water (Depending on the sensitivity of the GC-MS, a concentration of around 0.2 mg/mL to 2.0 mg/mL is ideal.).
5. Methoxyamine hydrochloride in pyridine (20 mg/mL).
6. *N*-Methyl-*N*-trimethylsilyltrifluoroacetamide (MSTFA).
7. 1.5 mL Eppendorf tube (*see* **Note 2**).
8. Mill (TUBE-MILL CS001, IKA).
9. −80 °C deep freezer.
10. Lyophilizer (FDU-1200, EYELA).
11. Vortex mixer (Micro tube Mixer, MT-400, TOMY).
12. Centrifuge (3520, KUBOTA).

13. Centrifugal evaporator (CVE-2000, EYELA).
14. Vials (include insert and septum) for the autosampler attached to the GC-MS.

3 Methods

3.1 Sample Preparation Before Extraction

1. Place approximately 1 g of sample (*see* **Note 3**) in a freeze-drying bottle and freeze at −80 °C.
2. After freezing, dry at room temperature overnight in a freeze dryer.
3. After drying, grind into a powder (homogenize).
4. Weigh 50 mg of the homogenized powder sample into a 1.5 mL Eppendorf tube (*see* **Note 3**).

3.2 Main Extraction Method

1. Add 1 mL of a solvent comprising methanol/chloroform/ultrapure water (2.5/1/1 = v/v/v) to the weighed samples.
2. Add an internal standard substance (e.g., 60 μL of ribitol (0.2 mg/mL to 2 mg/mL (*see* **Note 4**).
3. Mix for 5 min using a mixer and then centrifuge (16,000 *g* for 5 min at 4 °C).
4. Transfer 800 μL of the supernatant after centrifugation to a new Eppendorf tube and add 400 μL of ultrapure water.
5. After stirring for 1 min, centrifuge the suspension; then, transfer 400 μL of the supernatant to a new Eppendorf tube.
6. Concentrate using a centrifugal evaporator for 1 h.
7. Freeze at −80 °C and dry in a freeze dryer (*see* **Note 5**).
8. After drying, derivatize.

3.3 Water Extraction

1. Add 1 mL of ultrapure water and an internal standard (as in Subheading 3.2, **step 2**) to the weighed sample.
2. Mix for 5 min using a mixer, then centrifuge (16,000 *g* for 5 min at 4 °C).
3. Transfer the entire supernatant after centrifugation to a new Eppendorf tube.
4. Freeze at −80 °C and dry in a freeze dryer.
5. After drying, derivatize.

3.4 Extraction with 80% Ethanol

1. Add 1 mL 80% ethanol and the internal standard (as in Subheading 3.2, **step 2**) to the weighed sample.
2. Mix for 5 minutes using a mixer, then centrifuge (16,000 *g* for 5 min at 4 °C).
3. Transfer the entire supernatant after centrifugation to a new Eppendorf tube.

4. Concentrate using a centrifugal evaporator for 2 h.
5. Freeze at −80 °C and dry in a freeze dryer.
6. After drying, derivatize.

3.5 Derivatization

1. Add 50 μL of methoxyamine hydrochloride in pyridine to the Eppendorf tube after drying, stir for 1 min, and allow to react at 30 °C for 90 min.
2. Add 100 μL of MSTFA, stir for 1 min, and allow to react at 37 °C for 30 min.
3. Transfer the entire sample to a GC-MS vial and subject it to GC-MS analysis.

4 Notes

1. To visualize the component profiles obtained using the three extraction methods described in the methods section, score plots obtained using principal component analysis are shown in Fig. 1. The experimental conditions were as follows: Frozen mackerel meat was freeze-dried and ground into a powder, and the experiment was repeated three times as described in the Methods section. After GC-MS analysis, the detected peaks were annotated as previously described [5]. The annotated components are listed in a data table, and principal component analysis was performed using SIMCA 14. Consequently, the cumulative contribution ratios of the first and second principal

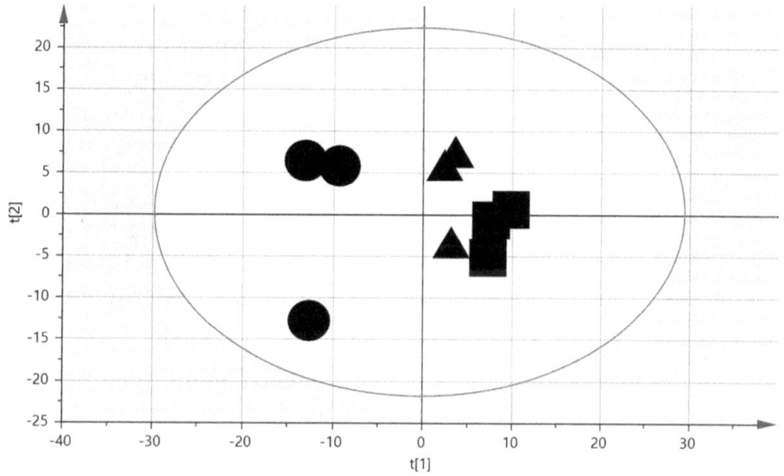

Fig. 1 Score plot obtained by principal component analysis (● Extraction with methanol/chloroform/ultrapure water (2.5/1/1 = v/v/v), ▲extraction with water, ■ extraction with 80% ethanol, t [1] First principal component: 48.4% contribution rate, t [2] second principal component: 15.9% contribution rate)

components were 64.3%. Based on the first principal component, the extraction method using methanol/chloroform/ultrapure water was positioned in the negative direction, whereas the water and 80% ethanol extractions were positioned in the positive direction. This indicates that the methanol/chloroform/ultrapure water method and the other two methods showed different component profiles. Therefore, differences in the extraction methods affected the data obtained from GC-MS metabolomics. As the choice of the extraction method affects the subsequent analysis, it is important to decide on the extraction method according to the experimental objective.

2. It is better to use the same type of 1.5 mL Eppendorf tube in the same experiment. Depending on the Eppendorf tube used, there is a possibility that the components obtained from the fish may differ due to adsorption and other conditions. The same can be said of the other reagents, equipment, and instruments.

The example given is of a muscle region; however, the same method can be used to analyze any part or tissue of a fish, in addition to the muscle, the gills, liver, and skin were tested (unpublished data). However, the components obtained differed depending on the parts used. For example, the component profiles of the tail, belly, and back meat of yellowtail were different [5].

Although this is not related to the food industry, it is possible to use blood samples to analyze the nutrition and health of fish. The extraction of metabolites from fish blood can be performed in the same way as for human blood. Methanol extraction has been reported as a method for extracting fish blood [11].

3. Regarding the amount of sample to be used: There is no need to limit the amount of sample used for the initial freeze-drying of the tissue to 1 g. The amount required to achieve uniformity should be determined based on the purpose of the experiment.

In our study, the amount of powdered sample used after freeze-drying was 50 mg. Considering the size of the Eppendorf tube and the amount of extraction solvent, 50 mg is a good amount, but it was possible to conduct experiments using amounts smaller than 50 mg.

4. There is no need to limit the internal standard to ribitol. In this study, we used ribitol; however, 2-isopropylmalic acid was also used. It is necessary to select an effective internal standard according to the purpose of the experiment.
5. When freeze-drying the extraction solvent, we recommend covering the Eppendorf tube with Parafilm and making a hole in it with a needle before proceeding. The reason for this is

that, depending on the experimental sample, there is a possibility that the powdered sample will become airborne as powder dust during the experiment. This is less likely to occur in fish samples that contain a large amount of fat, but we recommend using parafilm.

There are limits to the components that can be detected using a single extraction method or GC-MS analysis. Depending on the selected extraction method, it may not be possible to detect the target components. It may be difficult to exhaustively detect all components in a single extraction using current techniques. In such cases, rather than trying to find a way to extract specific components at once using the same extraction method, it may be a good idea to measure them using a different established method and then integrate the data with the non-target analysis data when analyzing the data.

References

1. Mabuchi R, Adachi M, Ishimaru A, Zhao H, Kikutani H, Tanimoto S (2019) Changes in metabolic profiles of Yellowtail (*Seriola quinqueradiata*) muscle during cold storage as a freshness evaluation tool based on GC-MS metabolomics. Foods 8:511. https://doi.org/10.3390/foods8100511
2. Mabuchi R, Ishimaru A, Tanaka M, Kawaguchi O, Tanimoto S (2018) Metabolic profiling of fish meat by GC-MS analysis, and correlations with taste attributes obtained using an electronic tongue. Metabolites 9:1. https://doi.org/10.3390/metabo9010001
3. Tanimoto S, Kikutani H, Kitabayashi K, Ohkita T, Arita R, Nishimura S, Takemoto R, Mabuchi R, Shimoda M (2018) Qualitative changes in each part of yellowtail *Seriola quinqueradiata* flesh during cold storage. Fish Sci 84:135–148
4. Nevigato T, Masci M, Orban E, Lena GD, Casini I, Caproni R (2012) Analysis of fatty acids in 12 Mediterranean fish species: advantages and limitations of a new GC-FID/GCS-based technique. Lipids 47:741–753. https://doi.org/10.1007/s11745-012-3679-9
5. Mabuchi R, Adachi M, Kikutani H, Tanimoto S (2018) Discriminant analysis of muscle tissue type in yellowtail *Seriola Quinqueradiata* muscle based on metabolic component profiles. Food Sci Technol Res 24:883–891
6. Mabuchi R, Ishimaru A, Adachi M, Zhao H, Kikutani H, Tanimoto S (2019) Taste evaluation of Yellowtail (*Seriola Quinqueradiata*) ordinary and dark muscle by metabolic profiling. Molecules 24:2574
7. Furuta A, Saba M, Mabuchi R, Tanimoto S (2025) Changes in low-molecular-weight compounds in different parts of yellowtail (*Seriola quinqueradiata*) muscle under various heating conditions. Food Sci Technol Res 31(1):65–73
8. Saifer A (1971) Comparative study of various extraction methods for the quantitative determination of free amino acids from brain tissue. Anal Biochem 40:412–423. https://doi.org/10.1016/0003-2697(71)90401-5
9. Murata M, Sakaguchi M, Shimizu T, Eguchi H (1990) Changes in flavor profile in boiled muscle extracts of Yellowtail *Seriola quinqueradiata* stored in ice. Nippon Suisan Gakkaishi 56:697. https://doi.org/10.2331/suisan.56.697
10. Putri SP, Ikram MMM, Sato A, Dahlan HA, Rahmawati D, Ohto Y, Fukusaki E (2022) Application of gas chromatography-mass spectrometry-based metabolomics in food science and technology. J Biosci Bioeng 133:425–435. https://doi.org/10.1016/j.jbiosc.2022.01.011
11. Guo C, Huang X-Y, Yang M-J, Wang S, Ren S-T, Li H, Peng X-X (2014) GC/MS-based metabolomics approach to identify biomarkers differentiating survivals from death in crucian carps infected by *Edwardsiella tarda*. Fish Shellfish Immunol 39:215–222. https://doi.org/10.1016/j.fsi.2014.04.017

Chapter 19

GC-MS-Based Metabolite Profiling Method for White Leg Shrimp (*Litopenaeus vannamei*)

Rafidha Irdiani and Safira Latifa Erlangga Putri

Abstract

Litopenaeus vannamei, commonly known as white leg shrimp, has dominated the global shrimp industry owing to its superior yield performance, high tolerance to disease, and fast production cycle. It is also highly demanded by consumers for its desirable flavor. To quantitatively assess the quality of white leg shrimp, gas chromatography/mass spectrometry (GC/MS) analysis is often used. It can detect both volatile and non-volatile compounds that influence shrimp's nutritional and sensory quality. Although many studies have employed GC-MS-based metabolite profiling on white leg shrimp, detailed guidance on conducting such analyses remains limited. On the other hand, such guidance is essential to support researchers who are new to this field and to aid in the development of quality assessment methods for white leg shrimp. Therefore, this chapter aims to provide a comprehensive guide on GC/MS analysis for white leg shrimp. In this chapter, a step-by-step procedure will be discussed, starting from sample preparation, extraction, and derivatization, until the injection of the samples into the GC-MS instruments.

Key words Gas chromatography, Mass spectrometry, Metabolite profiling, Metabolomics, Shrimp

1 Introduction

Litopenaeus vannamei, also known as white leg shrimp, has dominated the global shrimp industry owing to its superior yield performance, high tolerance to disease, fast production cycle, and high consumer demand [1]. Several countries in Asia, such as Indonesia, India, China, Vietnam, and Thailand, and several Latin-American countries have become the major contributors to the white leg shrimp production worldwide [2]. The production and consumption of white leg shrimp are consistently increasing each year [3].

White leg shrimp is renowned for its desirable flavor. Prior studies show that it is rich in free amino acids and flavor-active nucleotides, and that their concentrations vary with aquaculture practices and the shrimp's geographic origin [4–6]. From a consumer perspective, larger shrimp are perceived as more palatable and therefore they are priced higher than smaller one. Consistent

Sastia Prama Putri (ed.), *Gas Chromatography-Mass Spectrometry: Based Metabolomics in Food Science*, Methods and Protocols in Food Science, https://doi.org/10.1007/978-1-0716-5108-7_19,

with this perception, a quantitative descriptive analysis (QDA) revealed that larger shrimps were attributed to higher scores of sweetness, juiciness, and crispness—parameters associated with greater palatability [7].

Metabolomics is considered as one of the most effective approaches for understanding phenotypic changes in biological systems [8–10]. In food science, metabolomics enables both qualitative and quantitative evaluation of food samples due to the influence of various factors, such as processing methods, environments, and storage techniques. This leads to a rapid and reliable method for assessing food quality [9].

Gas chromatography-mass spectrometer (GC-MS) is one of the most widely used instruments to profile metabolites in food samples owing to its low cost, high reproducibility, and robust stability compared to other analytical instruments [11]. These advantages are due to the reliability of retention times and mass spectra produced by hard ionization [12]. The development of GC/MS analysis methods has enabled the analysis of both volatile and nonvolatile compounds with low molecular weight. In food, these compounds play essential roles in determining both nutritional and sensory quality [13, 14]. Numerous studies have applied GC-MS-based metabolite profiling to analyze a wide range of food samples, including legumes, fruits, coffee, cocoa beans, and fermented products [15–19].

In white leg shrimp, numerous studies employing GC-MS-based metabolite profiling have been conducted—providing valuable insights into potential sensory quality-promoting metabolites [7, 20–23]. However, detailed guidance on how to conduct a GC/MS analysis for white leg shrimp remains limited. In particular, no existing protocol addresses the precautions and considerations necessary for proper sample handling and preparation. This is essential to support researchers who are new to the field of metabolomics and to aid in the development of quality assessment methods for white leg shrimp. Therefore, this chapter aims to fill that gap by providing a detailed guide on GC/MS analysis for white leg shrimp. In this chapter, a step-by-step procedure will be discussed, starting from sample preparation, extraction, derivatization, and injection of the samples into the GC=MS instruments. This chapter focuses on the analysis of low molecular weight compounds using specific instruments and database references for annotation, as described in this protocol.

2 Materials

The materials required are categorized into three groups: samples and reagents, tools and laboratory supplies, and equipment and GC-MS-related essentials. All materials are listed in Table 1.

3 Methods

3.1 Sample Preparation

1. Disinfect the knife and cutting board with 70% ethanol using lint-free wipes.
2. Place the shrimps inside a styrofoam thermal box containing ice immediately after harvest.
3. Wash the shrimp, remove the head along the centerline, peel, and remove the digestive track of the shrimp.
4. Collect the remaining body tissues (muscle and gut), cut them into small pieces, and into 50 mL polycarbonate tubes.
5. Pour some liquid nitrogen into a styrofoam box and immerse the tubes for at least 2 min. Samples can be stored at −30°C before further processing. If stored, immerse the tubes in liquid nitrogen for 2 min before proceeding to the next step.
6. Remove the tubes from liquid nitrogen by using laboratory tongs and place them in a 50 mL tube rack.
7. Immerse the freeze-drying flask in liquid nitrogen while performing the following steps.
8. Remove the tube caps and seal the tubes with sealing film.
9. Puncture small holes in the sealing film by using a pin or awl to allow moisture to escape during freeze-drying.
10. Remove the freeze-drying flask from the liquid nitrogen and place the tubes inside it. Secure the flask with its rubber cap.
11. Attach the flask to the freeze dryer and dry for at least 14 h.
12. After 14 h, samples can be either stored at −30°C or directly ground.
13. To grind the samples, place a metal bead into each polycarbonate tube and seal it with the original cap.
14. Immerse the tubes in liquid nitrogen for 2 min.
15. Grind the samples using a multi-bead shocker at 2,500 rpm for 40 s.
16. Re-immerse the tubes in liquid nitrogen. If samples are not yet fully powdered, or small chunks remain, repeat the grinding, and immerse them again in liquid nitrogen afterward.
17. Allow the tubes to reach room temperature (approximately taken 30 min to 1 h). Then, remove the metal bead using a magnet.
18. Store the powdered samples at −30°C until further analysis.

3.2 Extraction

1. Place 5 mg of the powdered sample into a 2 mL microtube using a laboratory spatula.

Table 1
List of materials

No	Materials	Sample preparation	Extraction	Derivatization	GC-MS analysis
Samples and reagents					
1	White leg shrimp	✓			
2	70% ethanol disinfectant	✓	✓	✓	
3	Ethanol (HPLC[a] grade)		✓		
4	Ultrapure water		✓		
5	Ribitol		✓		
6	Methoxyamine hydrochloride			✓	
7	Infinity pure pyridine			✓	
8	N-methyl-N-(trimethylsilyl)-trifluoroacetamide (MSTFA)			✓	
9	Alkane standard mixture (C_8—C_{40})				✓
10	Pyridine				✓
11	Methanol				✓
12	Hexane				✓
Tools and laboratory supplies					
13	Nitrile gloves	✓	✓	✓	✓
14	Knife	✓			
15	Cutting board	✓			
16	50 mL polycarbonate tube	✓			
17	50 mL tube rack	✓			
18	Metal beads	✓			
19	Magnet	✓			
20	Cryogenic gloves	✓			
21	Laboratory tongs	✓			
22	Sealing film	✓			
23	Styrofoam thermal box	✓			
24	Freeze-drying flask	✓			
25	Ice	✓			
26	Liquid nitrogen	✓			
27	Laboratory permanent marker	✓	✓	✓	✓
28	Pin or awl	✓	✓		

(continued)

Table 1 (continued)

No	Materials	Sample preparation	Extraction	Derivatization	GC-MS analysis
29	Laboratory spatula		✓	✓	
30	Measuring glass		✓	✓	
31	Microtube (2 mL)		✓		
32	Micropipette		✓	✓	✓
33	Tips		✓	✓	✓
34	Microtube rack		✓	✓	✓
35	15 mL conical centrifuge tube		✓		
36	Laboratory glass bottle		✓	✓	
37	GC-MS vials equipped with 200 μL insert				✓
38	Lint-free wipe	✓			✓
Equipment					
39	Freeze dryer	✓			
40	Multi-beads shocker	✓			
41	Freezer (−20 °C or −30 °C)	✓			
42	Analytical balance		✓		
43	Refrigerator (4 °C)		✓		
44	Thermomixer		✓	✓	
45	Vortex mixer		✓		
46	Centrifuge		✓	✓	
47	Centrifugal concentrator		✓		
GC-MS-related essentials					
48	GC-MS system equipped with an autosampler and a syringe				✓
49	GC-MS inlet septum				✓
50	GC-MS inlet insert (liner)				✓
51	GC column: InertCap 5 MS/NP (GL sciences) or equivalent				✓
52	GC-MS library for annotation (NIST library and GL sciences DB)[b]				✓

[a]High-performance liquid chromatography

[b]Access https://systemsomicslab.github.io/compms/msdial/main.html to download GC Sciences DB library

2. Prepare 80% ethanol by mixing HPLC-grade ethanol and ultrapure water in a clean laboratory glass bottle.
3. Prepare a 0.25 mg/mL ribitol solution in ultrapure water to serve as an internal standard.

   The remaining ribitol stock can be stored at 4° C for future use.

4. Add 200 μL of ribitol solution to each 5 mg of sample and to empty tubes as blanks. Mix using vortex mixer for around 10 s, and homogenize using a thermomixer at 1,200 rpm, 25°C for 10 min.

   The number of blanks should correspond to the number of analytical batch. One analytical batch should have at least one blank (1 batch = ± 24 h).

5. Add 800 μL of 80% ethanol to both the sample and blank to achieve a final ribitol concentration of 50 μg/mL. Mix using a vortex mixer for 10 s.

   This solvent acts as a deproteinization agent.

6. Centrifuge the mixtures at 16,000 × g, 4°C for 10 min to allow protein precipitation.
7. Transfer 700 μL of the supernatant from each tube (sample and blank) into new microtubes.
8. From the 700 μL of supernatant of each sample, collect 100 μL into a 15 mL conical centrifuge tube to create a pooled sample for quality control (QC) in GC/MS analysis.

   The container size for the pooled sample may be adjusted depending on the number of samples.

9. Transfer another 100 μL of the remaining supernatant from both the sample and blank into new microtubes.

   The rest of the supernatant can be stored at − 30° C as backup.

10. Mix the pooled supernatant (QC) by using a vortex mixer and transfer 100 μL into a new microtube for use as QC samples.

    The number of QC samples should correspond to the duration of GC/MS analysis. It is recommended to inject one QC every 3–4 samples.

11. Replace the microtube caps temporarily with a perforated caps to allow solvent evaporation in the next step.

    Perforated caps can be made by cutting the caps from new microtubes and puncturing 3–4 holes using a pin or awl.

12. Dry the samples, blanks, and QCs tubes using a centrifugal concentrator for 10 min, followed by overnight lyophilization in a freeze dryer.

Alternatively, drying may be performed solely using a centrifugal concentrator. However, the duration must be optimized to ensure complete solvent removal.

13. Once all solvent has evaporated, remove the perforated caps and seal the tubes with their original caps.

Samples can be stored at −30°C until derivatization.

3.3 Derivatization

1. Prepare a 20 mg/mL methoxyamine hydrochloride solution in infinity pure pyridine.
2. Add 100 μL of methoxyamine hydrochloride solution to each dried sample extract, blank, and QCs. Incubate the mixture in a thermomixer at 30°C for 90 min.
3. Add 50 μL of MSTFA to each tube and incubate in a thermomixer at 37°C for 30 min.
4. Centrifuge the samples, blanks, and QCs at 16,000 × g, 25°C for 3 min to prevent contaminants from entering the GC column.
5. Transfer 100 μL of the mixture to GC-MS vials for analysis.

3.4 GC/MS Analysis

1. Prepare two 50 μL aliquots of alkane mix solutions in separate GC-MS vials.

It is recommended to inject alkane mix solutions at the beginning and end of each analytical batch for retention index calibration.

2. Prepare 100 μL of infinity pure pyridine in several GC-MS vials for column washing.

The number of vials should correspond to the number of samples. It is recommended to use each vial for no more than five times washing.

3. Clean the needle part of the GC syringe using lint-free wipes soaked with methanol, followed by hexane. To clean the inner part, aspirate and dispense each solvent separately at least 2–3 times.
4. Prepare the syringe washing tubes in the autosampler by adding 2–3 mL of pyridine to each tube.

The amount of pyridine should correspond to the number of analytical batch and instrument setting.

5. Perform a leak check and autotuning to ensure system readiness.

6. Place the GC-MS vials containing the samples, blanks, QCs, alkane mix, and pyridine for column washing into the autosampler rack.
7. Set the GC-MS parameters (Table 2, 3, and 4).

The GC-MS settings vary depending on the column type and target compounds. The following parameters are specific to the conditions used in this protocol.

8. Input the batch file.

Depending on the manufacturer, some information needs to be included, such as sample name, vial number, data file, method file, and tuning file

9. Start the analysis.
10. Perform peak processing and compound annotation.

For the conditions used in this protocol, the NIST and GL Sciences DB library can be used for annotation.

Table 2
GC/MS analysis parameters for samples

Parameters	Value
Injection volume	1 μL
Injection mode	Split mode of 25:1 (v/v) ratio
Injection temperature	230°C
Column temperature	Initially held at 80 °C for 2 min, followed by an increase by 15°C/min until reaching 330°C, and was maintained for 6 min
Transfer line temperature	250°C
Ion source	200°C
Ionization method	Electron ionization (EI)
Mass range	85–500 Da
Carrier gas	He with a column flow rate of 1.12 mL/min and a linear velocity of 39 cm/s

Table 3
GC/MS analysis parameters for alkane mix

Parameters	Value
Injection volume	1 μL
Injection mode	Split mode of 12:1 (v/v) ratio
Injection temperature	230°C
Column temperature	Initially held at 80°C for 2 min, followed by an increase by 15°C/min until reaching 330°C and was maintained for 12 min
Transfer line temperature	250°C
Ion source	200°C
Ionization method	Electron ionization (EI)
Mass range	85–500 Da
Carrier gas	He with a column flow rate of 1.12 mL/min and a linear velocity of 39 cm/s

Table 4
GC/MS analysis parameters for pyridine

Parameters	Value
Injection volume	1 μL
Injection mode	Split mode of 25:1 (v/v) ratio
Injection temperature	230°C
Column temperature	Initially held at 80°C for 1 min, followed by an increase by 30°C/min until reaching 330°C and was maintained for 3 min
Transfer line temperature	250°C
Ion source	200°C
Ionization method	Electron ionization (EI)
Mass range	85–500 Da
Carrier gas	He with a column flow rate of 1.12 mL/min and a linear velocity of 39 cm/s

4 Notes

For optimal results, the following precautions and considerations should be taken into account during sample preparation until GC/MS analysis.

4.1 Notes for Sample Preparation

1. During sample preparation, although temporary storage at −30°C is acceptable, frequent freeze-thaw cycles are not recommended if the samples have not yet been freeze-dried. It is best to complete the entire sample preparation at a time.
2. Ensure that the knife and cutting board are cleaned with 70% ethanol before processing shrimp with different characteristics to avoid cross-contamination.
3. For larger sized shrimp (headless body weight > 25 g), if the shrimp does not fit into a single 50 mL polycarbonate tube, divide it into two or more tubes. After grinding, combine all dried powders before weighing to ensure consistency.
4. Multiple 50 mL polycarbonate tubes can fit inside one freeze-drying flask. While placing the tubes inside the flask, make sure none of them obstruct the mouth of another. This is important to ensure that all samples are dry properly and at the same rate.

4.2 Notes for Extraction

1. Wait until the tubes reach room temperature before opening the caps for weighing. This aims to prevent moisture from condensing inside the tubes due to temperature changes.
2. Clean the laboratory spatula with methanol or ethanol before using it on samples with different characteristics to avoid cross-contamination.
3. It is important to ensure that no solvent remains after drying. Tap the bottom of the microtube to check for any residual droplets.

4.3 Notes for Derivatization

1. The derivatization efficiency of MSTFA is optimal within 24 h. If the GC/MS analysis for all samples takes longer than 24 h, it is best to divide the analysis into multiple batches.
2. A glass container should be used to store the methoxyamine hydrochloride solution.
3. Prepare the methoxyamine hydrochloride solution on the same day as the derivatization. If having multiple analytical batches, prepare a fresh solution for each batch.

4.4 Notes for GC/MS Analysis

1. It is recommended to frequently wash the column with pyridine between samples. Specifically, two injections of pyridine should be performed after every two samples and one QC injection.
2. Under the current protocol conditions, the inlet insert should be replaced after every 50 injections.

References

1. Dayalan V, Kasivelu G, Raguraman V, Sharma AN (2024) Studies on temperature impact (sudden and gradual) of the white-leg shrimp *Litopenaeus vannamei*. Environ Sci Pollut Res 31(27):38743–38750
2. Jamal MR (2023) Can vannamei shrimp (*Litopenaeus vannamei*) revitalise Bangladesh's dying shrimp industry? Aquac Int 31(3): 1637–1641
3. Kim S-H, Jung E-J, Hong D-L, Lee S-E, Lee Y-B, Cho S, Kim S-B (2020) Quality assessment and acceptability of whiteleg shrimp (*Litopenaeus vannamei*) using biochemical parameters. Fish Aquat Sci 23:1–10
4. Chang L-T, Fang M (2024) Establishment of the flavor quality standard for white shrimp (*Litopenaeus vannamei*) by instrumental and sensory approaches. ACS Food Sci Technol 4(12):3016–3024
5. Liang M, Wang S, Wang J, Chang Q, Mai K (2008) Comparison of flavor components in shrimp *Litopenaeus vannamei* cultured in sea water and low salinity water. Fish Sci 74:1173–1179
6. Hsieh H-H, Weerathunga V, Weerakkody WS, Huang W-J, Muller FLL, Benfield MC, Hung C-C (2021) The effects of low pH on the taste and amino acid composition of tiger shrimp. Sci Rep 11(1):21180
7. Putri SLE, Suantika G, Situmorang ML, Christina J, Nikijuluw C, Putri SP, Fukusaki E (2021) Shrimp count size: GC/MS-based metabolomics approach and quantitative descriptive analysis (QDA) reveal the importance of size in white leg shrimp (*Litopenaeus vannamei*). Metabolomics 17:1–12
8. Utpott M, Rodrigues E, de Oliveira Rios A, Mercali GD, Flôres SH (2022) Metabolomics: an analytical technique for food processing evaluation. Food Chem 366:130685
9. Wu W, Zhang L, Zheng X, Huang Q, Farag MA, Zhu R, Zhao C (2022) Emerging applications of metabolomics in food science and future trends. Food Chemistry: X 16:100500
10. Putri SP, Nakayama Y, Matsuda F, Uchikata T, Kobayashi S, Matsubara A, Fukusaki E (2013) Current metabolomics: practical applications. J Biosci Bioeng 115(6):579–589
11. Beale DJ, Pinu FR, Kouremenos KA, Poojary MM, Narayana VK, Boughton BA, Kanojia K, Dayalan S, Jones OAH, Dias DA (2018) Review of recent developments in GC–MS approaches to metabolomics-based research. Metabolomics 14:1–31
12. Putri SP, Ikram MMM, Sato A, Dahlan HA, Rahmawati D, Ohto Y, Fukusaki E (2022) Application of gas chromatography-mass spectrometry-based metabolomics in food science and technology. J Biosci Bioeng 133(5): 425–435
13. Pedrosa MC, Lima L, Heleno S, Carocho M, Ferreira ICFR, Barros L (2021) Food metabolites as tools for authentication, processing, and nutritive value assessment. Foods 10(9):2213
14. Sato F, Matsui K (2012) Engineering the biosynthesis of low molecular weight metabolites for quality traits (essential nutrients, health-promoting phytochemicals, volatiles, and aroma compounds). In: Plant biotechnology and agriculture. Elsevier, pp 443–461
15. Amalia F, Irifune T, Takegami T, Yusianto, Sumirat U, Putri SP, Fukusaki E (2023) Identification of potential quality markers in Indonesia's Arabica specialty coffee using GC/MS-based metabolomics approach. Metabolomics 19(11):90
16. Dahlan HA, Nambu Y, Putri SP, Fukusaki E (2022) Effects of soaking Tempe in vinegar on metabolome and sensory profiles. Metabolites 12(1):30
17. Hanifah A, Firmanto H, Putri SP, Fukusaki E (2022) Unique metabolite profiles of Indonesian cocoa beans from different origins and their correlation with temperature. J Biosci Bioeng 134(2):125–132
18. Iman MN, Irdiani R, Rahmawati D, Fukusaki E, Putri SP (2023) Improvement of the functional value of green soybean (edamame) using germination and Tempe fermentation: a comparative metabolomics study. J Biosci Bioeng 136(3):205–212
19. Yunindanova MB, Putri SP, Novarianto H, Fukusaki E (2024) Characteristics of kopyor coconut (*Cocos nucifera* L.) using sensory analysis and metabolomics-based approach. J Biosci Bioeng 138(1):44–53
20. Putri SLE, Suantika G, Situmorang ML, Putri SP, Fukusaki E (2022) Metabolomics approach to elucidate the importance of count size in commercial penaeid shrimps: white leg shrimp (*Litopenaeus vannamei*) and black tiger shrimp (*Penaeus monodon*). J Biosci Bioeng 133(5): 459–466

21. Shi J, Xiao N, Zhang Q, Tian Z, Li M, Shi W (2024) Evaluation of aroma characteristics of *Penaeus vannamei* with different drying methods using HS-SPME-GC-MS, MMSE-GC-MS, and sensory evaluation. Food Chem 449: 138957
22. Suantika G, Situmorang ML, Saputra FI, Putri SLE, Putri SP, Aditiawati P, Fukusaki E (2020) Metabolite profiling of whiteleg shrimp *Litopenaeus vannamei* from super-intensive culture in closed aquaculture systems: a recirculating aquaculture system and a hybrid zero water discharge–recirculating aquaculture system. Metabolomics 16:1–11
23. Zhang D, Ji H-W, Luo G-X, Chen H, Liu S-C, Mao W-J (2021) Insight into aroma attributes change during the hot-air-drying process of white shrimp using GC-MS, E-nose and sensory analysis. Food Sci Technol 42:e70820

Chapter 20

GC-MS-Based Metabolite Profiling Method for Terasi

Rifqi Ahmad Riyanto

Abstract

Terasi, a traditional Indonesian shrimp paste, is renowned for its rich and complex flavors, which arise from a diverse array of compounds. To enhance our understanding of terasi's quality, it is essential to analyze these compounds systematically. Employing a metabolomics approach offers a detailed metabolite profiling that captures the full spectrum of substances present in terasi. This chapter provides an in-depth overview of the gas chromatography-mass spectrometry (GC-MS) workflow, a powerful technique used to identify and quantify the various compounds within the shrimp paste, thereby contributing valuable insights into its quality and flavor characteristics.

Key words Derivatization, Extraction, GC-MS, Nonvolatile analysis, Terasi

1 Introduction

In Indonesia, the traditional fermented shrimp paste is referred to as terasi. This product is mainly derived from planktonic shrimp species, particularly *Schizopodes* spp. and *Mydes* spp. Terasi serves as a condiment, enhancing dishes with both salty and umami flavors, thereby contributing to the overall culinary experience. Fermented shrimp paste is characterized by its rich color, which can range from dark brown to gray or red, and it is known for its distinct aroma. This ingredient is commonly combined with chili, garlic, and salt to create *sambal terasi*, a flavorful condiment that enhances various dishes [1].

The production of terasi involves four essential processing steps: (a) salting, (b) drying, (c) kneading, and (d) fermenting. Initially, approximately 10% (w/w) salt is incorporated with oceanic planktonic shrimp. This addition of salt serves to enhance the flavor profile while also extending the shelf life of the finished product. Then, the mixture is dried to a final moisture content of 50%, which occurs either naturally, where the mixture is spread on straw mats and sun-dried, or mechanically. Following the initial drying process, additional salt may occasionally be incorporated. The semidry

Sastia Prama Putri (ed.), *Gas Chromatography-Mass Spectrometry: Based Metabolomics in Food Science*, Methods and Protocols in Food Science, https://doi.org/10.1007/978-1-0716-5108-7_20,

shrimp mixture is then kneaded to ensure uniformity and is compactly pressed into wooden tubs to eliminate excess air. The mixture undergoes fermentation in the tubs for a duration ranging from a few weeks to several months. Once the fermentation process is complete, the product is subjected to further drying and subsequently packaged for market distribution [2].

The fermentation of shrimp is characterized by significant changes in bacterial populations. Initially, pathogenic bacteria, including *Vibrio*, *Photobacterium*, *Psychrobacter*, *Pseudoalteromonas*, and *Enterovibrio*, predominate during the early stages of the fermentation process. However, it is important to note that their populations diminish by the conclusion of fermentation while the spoilage bacteria *Halanaerobium* increased [3]. At the end of shrimp fermentation, there is a notable increase in various metabolites, including amino acids, acetic acid, butyric acid, and trimethylamine, which arise from the proliferation of spoilage bacteria. Furthermore, the bacteria present in shrimp paste produce several enzymes—such as proteases and aminopeptidases—that facilitate the generation of free amino acids. Additionally, lipases contribute to the formation of fatty acids [4].

In this chapter, we will present detailed methods for GC/MS-based analysis aimed at thoroughly assessing the low molecular hydrophilic compounds in terasi samples. This section will cover techniques for sample preparation, extraction, and derivatization. The use of GC/MS enables reliable annotation and identification of compounds, providing a deeper understanding of the chemical changes that occur during various processing stages. This technology not only identifies major compounds but also detects minor compounds that can significantly influence the quality of terasi. The insights gained from GC/MS analysis will help optimize terasi processing methods and enhance the overall quality of the final product.

2 Materials

2.1 Common Materials

1. Semi-wet terasi.
2. Powdered terasi.

2.2 Non-targeted GC/MS Analysis

2.2.1 Chemicals

1. Methanol (80%).
2. Ultrapure water.
3. Ribitol standard.
4. Methoxyamine hydrochloride.
5. Pyridine (99.5%).
6. N-Methyl-N-(trimethylsilyl) trifluoroacetamide (MSTFA).

7. Alkane mixture (C_9–C_{40}) standard.
8. Liquid nitrogen.

2.2.2 Laboratory Supplies

1. Grinding tube 50 mL.
2. Metal cone.
3. Eppendorf tube 2 mL.
4. Eppendorf tube 1.5 mL.
5. Eppendorf tube 15 mL.
6. GC/MS vials with cap.
7. Amber glass vial/bottle 100 mL.
8. Micropipette tips.
9. InertCap 5MS/NP (30 mm × 0.25 mm I.D., df. = 0.25 μm; GL Sciences, Japan).

2.2.3 Laboratory Equipment and Instruments

1. Micropipette.
2. Spatula.
3. Cryotank.
4. Multi-beads shocker.
5. Thermomixer.
6. Vortex.
7. Centrifuge.
8. Vacuum evaporator.
9. Freeze dryer.
10. GCMSQP2010 Ultra (Shimadzu, Kyoto, Japan).
11. GCMS-TQ8030 (Shimadzu, Kyoto, Japan).
12. AOC-20i/s autosampler (Shimadzu, Kyoto, Japan).

2.2.4 Software for Data Analysis

1. GCMS Solution software package (Shimadzu, Kyoto, Japan).
2. Reifycs Abf Converter (https://www.reifycs.com/AbfConverter/).
3. MS-DIAL.
4. GL-Science DB Library (InertCap 5MS-NP, Kovats RI, 494 records; http://prime.psc.riken.jp/Metabolomics_Software/MS-DIAL/GCMS%20DB_InertCap%205MS-NP_GLscience.msp).
5. SIMCA-P.
6. R Studio.

3 Methods

3.1 Nontargeted GC/MS Analysis

The GC-MS analysis was conducted according to Helmi et al. [4] and Sato et al. [5].

3.1.1 Powdered Terasi Preparation

1. Terasi samples were lyophilized using a freeze-drying machine overnight.
2. Put the lyophilized terasi sample into a 50 mL grinding tube (Yasui Kikai Co.).
3. Insert the clean metal cone (Yasui Kikai Co.) into the grinding tube.
4. Close the tube tightly.
5. Quench the tube, with terasi and metal cone inside, by dipping it in liquid nitrogen.
6. Put the quenched tube in a Multi-beads shocker (Yasui Kikai Co., Osaka, Japan). Make sure the tube is securely positioned and close the lid as per the specified machine instructions.
7. Shook the tube at 2000 rpm for 30 s.
8. Open the lid and check the grinding result inside the tube. It is recommended to repeat the previous **steps 3–4** times to produce fine powder.

3.1.2 Extraction Method

1. Transfer 10 mg of powdered terasi into a 2 mL Eppendorf tube. It is recommended to prepare triplicates ($n = 3$) for each sample.
2. Prepare an empty tube as a blank, and proceed with all the steps below with samples.
3. Add 1 mL of 80% methanol containing an internal standard (ribitol 0.2 mg/mL) into the tube.
4. Homogenize and incubate the mixture at 161× *g* for 20 min at 37 °C.
5. Centrifuge the mixture at 11,740× *g* for 10 min at 4 °C.
6. After mixing with vortex, the mixture was centrifuged at 11,740× *g* for 3 min at 4 °C.
7. Transfer 200 μL of supernatant into a new 1.5 mL tube.
8. The supernatant was centrifuged at 3× *g* for 90 min at 25 °C.
9. The extract was dried under vacuum overnight (10–12 h) and freeze-dried until completely dried.

3.1.3 Derivatization Method

1. Add 100 μL methoxyamine hydrochloride (20 mg/mL in pyridine) to lyophilized extract.
2. Incubate and homogenize the mixture at 161× *g* at 30 °C for 90 min.

3. Add 50 μL of MSTFA to the mixture.
4. Incubate and homogenize the mixture again at 161 × *g* at 37 °C for 30 min.
5. The derivatized sample and blank are ready to be injected into the GC-MS.

3.1.4 Gas Chromatography-Mass Spectrometry Analysis

1. GC/MS analysis is conducted using either GCMSQP2010 Ultra (Shimadzu, Kyoto, Japan) or GCMS-TQ8030 (Shimadzu, Kyoto, Japan) equipped with an autosampler AOC-20i/s (Shimadzu, Kyoto, Japan).
2. Set the injection mode to split mode (25:1 (v/v)).
3. Set the injection temperature to 230 °C.
4. Use helium as carrier gas with a flow rate of 1.12 mL/min and set the linear velocity at 39 cm/s.
5. Set the column temperature at 80 °C for 2 min, increased by 15 °C/min to reach 330 °C, and finally kept at 330 °C for 6 min.
6. Set transfer line temperature to 250 °C.
7. Set the ion source temperature to 200 °C.
8. Set the electron ionization at 0.93 kV.
9. Set the recorded EI scan at 10,000 u/sec with the mass range *m/z* 85–500.
10. Tune and calibrate the mass spectrometer before sample analysis.
11. Inject the standard alkane mixture (C_9–C_{40}) at the beginning of analysis for retention index calculation.
12. Inject 1 μL the derivatized samples and blank(s) randomly.

3.1.5 Data Preprocessing and Analysis

1. Convert the chromatogram to netCDF format using the GCMS Solution software package (Shimadzu, Kyoto, Japan).
2. Convert the netCDF file to ABF format using Reifycs Abf Converter.
3. Use ABF format files for peak detection, deconvolution analysis, identification, and alignment using MS-DIAL [6].
4. Set the peak detection parameter with a minimum height of 1000 amplitude.
5. Use a linear weight moving average smoothing method.
6. Set the smoothing level to five scans, with an average peak width of 20 scans.
7. Set the σ-window value to 0.5.
8. Set the EI spectra cut-off to 200 amplitudes for the deconvolution analysis.

9. Use the retention time of standard alkane (C_9—C_{40}) and GL-Science DB Library (InertCap 5MS-NP, Kovats RI,494 records) for peak identification.
10. Set the parameter for peak identification, annotation, and alignment as follows:
    (a) Retention index (RI) tolerance: 10.
    (b) Retention time (RT) tolerance: 0.5 min.
    (c) *m/z* tolerance: 0.5 Da.
    (d) EI similarity cutoff: 70%.
    (e) Identification score: 70%.
11. Normalize the height of the detected and annotated peak using the height of the internal standard peak.
12. Export the data matrix consisted of the raw and normalized peak height of the detected and annotated peaks.
13. Filter out the detected peaks by excluding the peaks that were not of biological origin by referring to the blank.
14. Filter out the detected peaks with a relative standard deviation of more than 30% based on the raw peak heights obtained from the QC.
15. The data matrix with filtered peaks will be used for further multivariate analysis such as PCA, OPLS-R, or OPLS-DA using SIMCA-P or R Studio.

References

1. Hajeb P, Jinap S (2012) Fermented shrimp products as source of umami in Southeast Asia. J Nutr Food Sci 01(S10). https://doi.org/10.4172/2155-9600.s10-006
2. Surono IS, Hosono A (1994) Chemical and aerobic bacterial composition of "Terasi", a traditional fermented product from Indonesia. J Food Hyg Soc Jpn 35(3):299. https://doi.org/10.3358/shokueishi.35.299
3. Lee SH, Jung JY, Jeon CO (2014) Microbial successions and metabolite changes during fermentation of salted shrimp (saeu-jeot) with different salt concentrations. PLoS One 9(2):1–12. https://doi.org/10.1371/journal.pone.0090115
4. Helmi H, Astuti DI, Putri SP, Sato A, Laviña WA, Fukusaki E, Aditiawati P (2022) Dynamic changes in the bacterial community and metabolic profile during fermentation of low-salt shrimp paste (Terasi). Metabolites 12(2). https://doi.org/10.3390/metabo12020118
5. Sato A, Astuti DI, Putri SP, Fukusaki E (2020) Quality improvement of semi-wet terasi by optimizing the starter culture ratio of controlled fermentation. HAYATI J Biosci 27(4): 320–329. https://doi.org/10.4308/hjb.27.4.320
6. Lai Z, Tsugawa H, Wohlgemuth G, Mehta S, Mueller M, Zheng Y, Ogiwara A, Meissen J, Showalter M, Takeuchi K, Kind T, Beal P, Arita M, Fiehn O (2018) Identifying metabolites by integrating metabolome databases with mass spectrometry cheminformatics. Nat Methods 15(1):53–56. https://doi.org/10.1038/nmeth.4512

INDEX

Sastia Prama Putri (ed.), *Gas Chromatography-Mass Spectrometry: Based Metabolomics in Food Science*, Methods and Protocols in Food Science, https://doi.org/10.1007/978-1-0716-5108-7,

J

L

M

MIX
Papier aus verantwortungsvollen Quellen
Paper from responsible sources
FSC® C105338

If you have any concerns about our products,
you can contact us on
ProductSafety@springernature.com

In case Publisher is established outside the EU,
the EU authorized representative is:
Springer Nature Customer Service Center GmbH
Europaplatz 3, 69115 Heidelberg, Germany

Printed by Libri Plureos GmbH
in Hamburg, Germany